W9-BLB-468

Paper
and
Paperboard
Manufacturing
and
Converting
Fundamentals

Also published by Miller Freeman Publications for the pulp and paper industry:

Modern Mechanical Pulping in the Pulp and Paper Industry
edited by Ken L. Patrick

Process Control for Pulp and Paper Mills
edited by Kenneth E. Smith

Trends and Developments in Papermaking
edited by John C. W. Evans

Maintenance Practices in Today's Paper Industry
edited by Ken L. Patrick

Modern Paper Finishing
edited by John C. W. Evans

Mastering Management
by Roberta Bhasin

Pulp and Paper North American Factbook

Lockwood-Post's Directory

Pulp and Paper Mill Map

Pulp Technology and Treatment for Paper
By James d'A. Clark

Pulp and Paper Dictionary
by John R. Lavigne

An Introduction to Paper Industry Instrumentation
by John R. Lavigne

Instrumentation Applications for the Pulp and Paper Industry
by John R. Lavigne

International Glossary of Technical Terms for the Pulp and Paper Industry
edited by Paul D. Van Derveer and Leonard E. Haas

International Pulp & Paper Directory

Pulp & Paper International Factbook

PPI Map of the European Pulp and Paper Industry

PPI Map of the Asian Pacific Rim's Pulp & Paper Mills

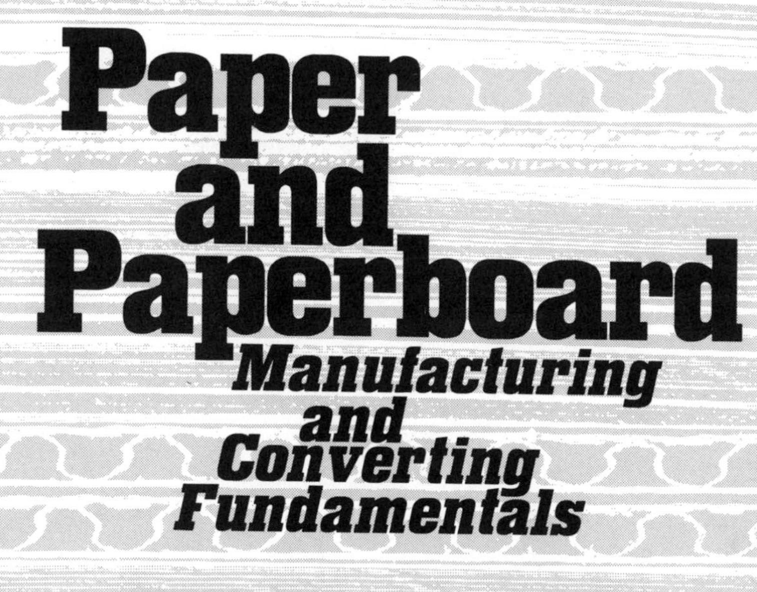

SECOND EDITION

by James E. Kline

MILLER FREEMAN PUBLICATIONS, INC.
San Francisco, Atlanta, Boston, Chicago, Cupertino,
New York, Brussels

A Pulp & Paper Book

Library of Congress Catalog Card Number: 91-61732
International Standard Book Number: 0-87930-190-2

Printed in the United States of America.

95 96 5 4 3 2

To Carolyn, Cathy, Jim and Christy

Contents

Preface

Every author feels his book should be everything to everybody: This one really is! Started as an introductory text for students in the Paper Science and Engineering curriculum at Western Michigan University, it soon expanded beyond that narrow scope. Frequent calls from friends, alumni and strangers all looking for some form of basic introductory book with some solid information led to an expansion of goals. The book covers all of the papermaking operations, combining material that describes the equipment with theoretical consideration of what is actually happening to the fibers. As such it should be of value to technical trainees as well as college students. By sprinkling discussion of paper and paperboard properties and how they are affected by each of the operations along the way, the book should be helpful to those selling and using paper. Further help to the manufacturing and selling group will be the section on converting operations. Last but certainly not least are the large family of suppliers who need to know what the industry is going to do with their products, and why we expect so much from them. The book is, in short, designed for anyone who wants to begin learning about the paper industry. It certainly is not the total story, but it should get you started.

The need for a second edition comes not from changes in the fundamentals, but from a need to adjust to current trends. Developments in such areas as recycling and computer control have gone beyond the coverage in the first edition. Some of the machinery and procedures that were new then are either gone or no longer need to be treated as experimental. New developments in the industry will continue, but the fundamental theories of operation will remain the same for many years to come.

ACKNOWLEDGMENT

To attempt to single out any one or two people who should be mentioned brings the risk of forgetting or neglecting someone. Being an "absent-minded professor," I have forgotten half of what I knew and most of where I learned it. However, this book would have been impossible without the help, advice, training and encouragement of my many associates and friends both in academia and the industry, but especially my early bosses and mentors, some of whom are not with us any more.

James E. Kline
Kalamazoo, Michigan

Section I

Introduction to the Paper and Paperboard Industry

The objective of this first section is to acquaint the reader with the basic design of the industry and the grades of paper and paperboard produced. The chapters are structured to present, first, some historical and economic background and a feel for the future potential for the industry. This is followed by a breakdown of the papermaking process into unit operations. Then, because the manufacturing operations are an important determinant of product properties, a discussion of those properties and how they are influenced by the manufacturing operations follows. This first section serves primarily as an introduction to the industry and also as an introduction and guide to the rest of the book:

1 Industry Overview

HISTORY AND DEVELOPMENT OF THE INDUSTRY

Literally thousands of different types of paper and paperboard are made. These products are so common that we use many of them without recognizing their source. When we do, it is usually because we want a disposable item, or something to carry a valuable message or work of art. Yet it is this same versatility, availability and disposability that make paper so important to our civilization and to the standard of living that we and other industrialized nations now enjoy.

The first use of paper is not accurately recorded, and therefore is credited to several different cultures. Ceremonial paper resembling cloth is alleged to have been made thousands of years ago in South and Central America. There are still cultures in the South Pacific where similar papers are made by beating bark with stones or logs. It has also been claimed that the Hausteca or Mayan Indians in Mexico made paper by suspending fibers in water and floating them onto a cloth. Because of better record keeping and the ability to establish a direct historical line, credit for the invention of papermaking is given to Tsai Lun, a Chinese Minister of Agriculture, who beat silk and mulberry bark together and screened the fibers from water with a bamboo mold. This invention in 105 A.D. is now heralded as the first time the present method of manufacture was used. The basic technique was refined by the Chinese and kept as a well-guarded secret until the eighth century, when it is reported to have been extracted from a prisoner of war and used in Samarkand.

The art of papermaking then spread through Central Asia, Asia Minor and Egypt and into Europe, where it was quite well established by 1400. During this period, the basic technique remained relatively unchanged. Fibers from many different sources were separated and suspended in a vat of water, and a mold or screen of some sort dipped into the vat and lifted out, separating the fibers from the water. After the sheet of paper was formed, it was pressed between felts and either hung or placed on a smooth surface to dry. This technique is still practiced in many parts of the world, primarily as an art form. A similar technique is practiced in Japan, where the fibers are suspended in a highly viscous water suspension, the sheet is formed by dipping the mold several times to build up the desired thickness, and the sheets are pressed without felts and then dried on a smooth surface. This technique has not yet been adapted to mechanical methods.

With the development of movable type by Gutenberg and the use of this invention to print the Bible in about 1450, literacy increased and the demand for paper soon outstripped the ability of men to make it fast enough by hand. Many developments increased the production rate of papermaking, but the most important would have to be the invention of papermaking machines around 1800. From that time to the present, the same techniques have been refined, streamlined, polished and made more efficient, but not substantially changed from Tsai Lun's original concept.

The development of the industry as outlined by these events can be seen to closely parallel the development of Western civilization. Paper has become an integral part of the development of our culture, both as a communications medium and in packaging. The per capita consumption figures shown in Table 1.1 reflect the relationship between paper use and industrial development in other cultures. Anyone who has traveled in the nations with per capita consumption levels below 200 lb/yr has noticed the absence of paper products normally taken for granted. As the less industrialized nations become more developed, the demand for paper is expected to grow, creating a continually increasing demand for paper and excellent prospects for long-term growth of the industry.

GRADE STRUCTURES

Because of the large number of grades made, it is difficult to group and present the grade structure in a brief and concise table. The best effort seems to be the breakdown presented in Table 1.2. The initial division is made on the basis of the weight of the product, calling the two categories thus distinguished *paper* and *paperboard*. Unfortunately, some paperboard grades are lighter or thinner than certain of the paper grades and there is no specific weight cut-off that separates the two categories. The division is also based on use—with the paper grades being used primarily for communications and the paperboard grades primarily for packaging. This method of division is also less than perfect, but by combining the two methods, the division becomes more understandable.

Many of the grades can be easily recognized, but the use of special terms by the industry makes others obscure. For example, *bristols* are heavy paper, and *glassine* and *creped* refer to specific manufacturing operations. The paperboard grades listed as *milk carton*, *cup stock*, *plate*, and so forth, are also known in the industry as either *solid bleached sulfate* or *fourdrinier board*. The listed grade *combination bending board* is better known to the consumer as cereal boxes or other boxes found in the grocery store, except for all-white cartons, which fall in the previous category. Corrugated cartons are made from linerboard on the outside and 9-point board or corrugating medium as the fluted layer. Although both look like unbleached kraft paper, they are categorized as paperboard. The category called *construction paper and board* includes some, but not all, of the products used in construction. Paper used with fiberglass insulating batts is an *unbleached kraft*, which leaves the mill as a roll of paper and is converted later. This leads to another source of confusion in classifying paper and paperboard grades: many grades are used for multiple purpose or are subjected to more than one conversion operation.

Because of this diversity of grades, it is not feasible to describe the manufacturing processes by grade. Furthermore, since many of the same operations are used in all of the different grades, structuring the discussion by grade would lead to substantial duplication. The book is therefore organized on the basis of the

Table 1.2. *Continued*

Type of grade	1988	1980
	(000 short tons)	
PAPERBOARD, TOTAL	**37,730**	**31,143**
Solid woodpulp furnish paperboard	**28,463**	**23,901**
Unbleached kraft packaging, industrial converting	18,384	15,285
Unbleached linerboard	16,988	14,249
Tube, can, and drum paperboard	68	220
Other, including corrugating medium	1,328	(D)
Bleached packaging, industrial converting (80% bleached fiber)	4,862	4,034
Linerboard	524	(D)
Folding carton-type board	2,063	2,063
Milk carton board	1,111	1,023
Heavyweight cup stock	566	386
Plate, dish, and tray stock	267	273
Other bleached board	331	111
Semichemical, including corrugating medium, other	5,222	4,583
Recycled paperboard	**9,162**	**7,242**
Recycled shipping containerboard	3,186	2,070
Linerboard	976	518
Corrugating medium	2,061	1,281
Container chip and filler board	149	271
Recycled folding carton board	2,550	2,350
Unlined chipboard	546	606
Kraft-lined	137	81
White-lined	117	220
Clay-coated	1,751	1,443
Recycled setup paperboard	302	560
Recycled packaging, industrial converting board	3,124	2,262
Tube, can, and drum stock	1,145	564
Gypsum linerboard	1,239	1,036
Panelboard and wallboard stock	35	113
Other	704	548
Wet machine board	**101**	**138**
Construction paper and board	**2,583**	**4,390**
Construction paper	356	1,369
Structural insulation board	2,227	3,021
ALL GRADES, TOTAL	**75,424**	**61,307**

Source: Bureau of the Census, as given in *Pulp & Paper 1989 North American Factbook*, pp. 126–129, 208–209.

Note: Categories shown are those now used by Bureau of the Census. Categories in older reports do not always correspond to the present ones.
(D) Data withheld to avoid disclosing figures for individual companies.

manufacturing operations, with accompanying information on the importance of these operations to the final properties of the paper or paperboard and on the grades produced by each operation.

MANUFACTURING OPERATIONS

Although the paper industry has developed into a complex industry able to produce a wide variety of products in many different ways, there is still an underlying similarity behind all of the different products. All paper and paperboard manufacturing processes are based on the same techniques and operations. Regardless of the form of the end product, certain operations or processes must be performed on the raw material to convert it into a final product. These basic operations are common to all products, but are modified to produce either slight or enormous differences in the final product. Futhermore, all of the products are made from practically the same raw material: the cellulosic fibers found in trees.

Figure 1.1. Papermaking basics

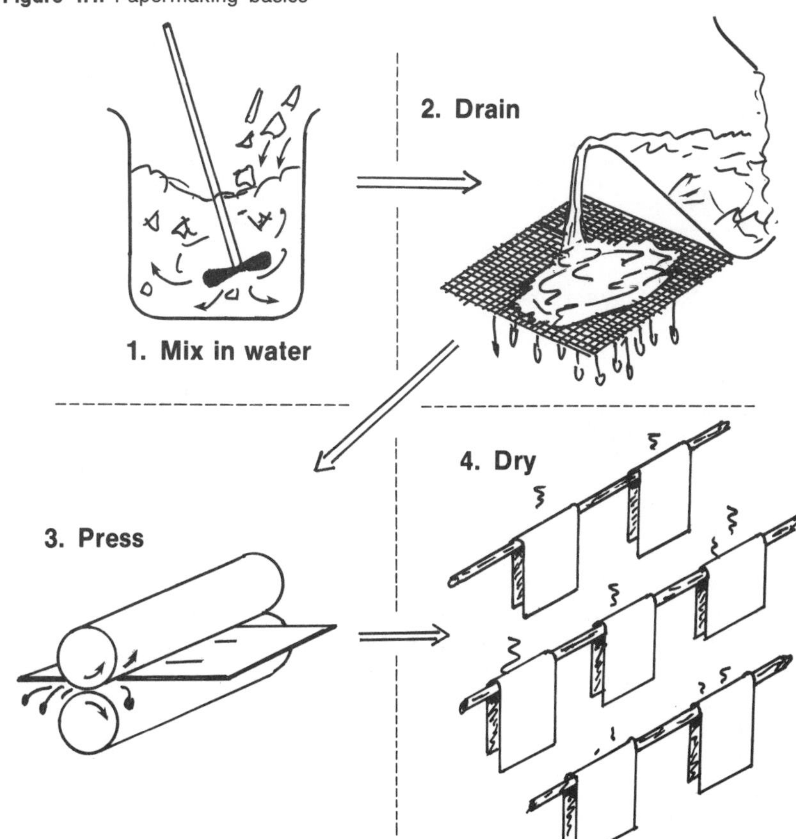

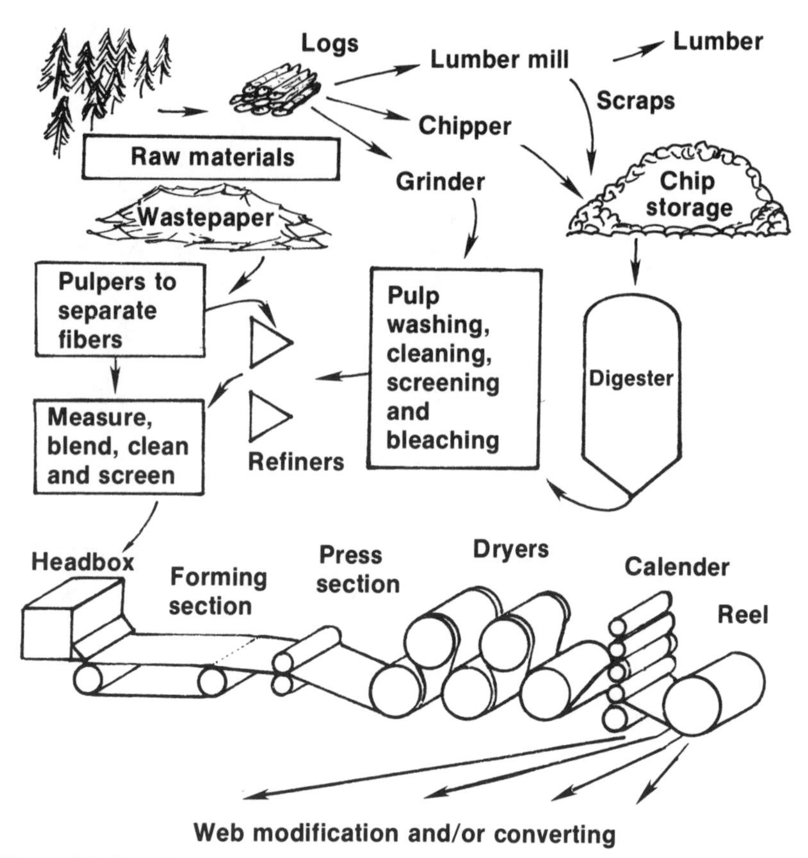

Figure 1.2. Overview of papermaking operations

We will see in later chapters that there are differences in the fibers found in different types of trees, but the differences are small compared with the differences in the properties of the products produced. Tissue, boxboard, printing paper, wallpaper, bags and sacks, freezer wrap and roofing felt can all be made from the same fibrous raw material. Furthermore, the fibers will go through essentially the same processes to become each of these diverse products. The differences between products are the result of the way in which the fibers are treated and the manufacturing process used in making each one.

The basic papermaking operations are summarized in Figure 1.1. All paper products are formed from fibers, which must first be removed from the raw material being used and separated into individual fibers. This process is called simply *mix* in Figure 1.1 but can vary from mixing to complex pulping operations. Once the fibers have been separated, they are formed into a random mat by *draining* the fiber-water mixture on some form of screen. The mat is then *pressed* and *dried* to complete the transformation into paper. While following these simple steps may suffice for the making of paper in the kitchen, it obviously is too simple

Table 1.3. Unit Operations of the Paper and Paperboard Industry

Pulping	Stock Preparation	Papermaking	Web Modification	Converting
• Selection and preparation of raw materials • Liberation of fibers	• Fiber modification • The furnish	• Preforming considerations • Forming devices • Consolidation of the web • Drying	• Surface modification • Physical modifications	• Printing • Corrugating • Packaging • Cut size and office supply • Tissue and related

to account for all of the grades listed in Table 1.2.

A better representation of the actual industry is made in Figure 1.2. The initial material may be anything from logs to wastepaper. The nature of the raw material and the properties of the final product determine which operations must be used to make each product. From this point to the end of the process, every step will have an impact on the final properties of the product. Although some operations may be common to several grades, and the final form of the product may not be determined until later, each step in the process remains important to the final product.

The operations shown in Figure 1.2 are also described in Table 1.3. The industry can be divided into a series of sequential unit operations that lead to the formation of the product. Within each operation there are several parallel routes that may be taken. For example, the first unit operation in Table 1.3, pulping, deals with the liberation of the fibers. In Figure 1.2 we can see that logs may go to the sawmill, the chipper or the grinder. Each will produce separated fibers, but the quality of the fibers from each one will be different. It can also be seen that the different routes require different degrees of treatment. Groundwood pulp, made by simply grinding logs, may go directly to the paper machine while chemical pulps generally require more treatment.

The second unit operation, stock preparation, further readies the fibers for their role in papermaking. Refiners modify the fibers physically and are the major factor in the development of strength in the paper. Blending of different fibers and the addition of chemicals are also included in the area of stock preparation.

The next operation, the actual paper or paperboard manufacturing, is divided into at least three suboperations: forming, pressing and drying. Many different forms of machines are used to produce the different grades of paper, and they will be considered later. At this time and for the purpose of Figure 1.2, the fourdrinier machine, which is used for the manufacture of most grades of paper, is illustrated.

The next set of operations listed in Table 1.3, web modification, is generally found or performed on the paper machine, but is easier to treat as a separate unit operation. Some machines may include none of these modification operations while others may have several. The most common of these operations improve the quality of the paper for printing by making the surface smoother or more resistant to water or ink.

The unit operations can be seen to progress from the initial selection of the raw material through breakdown of the raw material into fibers and then reassembly of the fibers into some product. There are other ways to group or organize these operations, but the listing used here follows the typical breakdown found in the industry. It is common for some of the operations to be performed by separate manufacturing units or even by different companies. Some plants exist primarily to produce pulp, performing only the operations shown in the first column of Table 1.3; however, most pulping operations are operated in conjunction with a papermaking operation. The pulp plant may also produce excess pulp, which it will sell to paper mills some distance from the forests. It is common for the pulp mill to be near the trees, but the paper mill need only be near water and shipping. The second and third unit operations, stock preparation and papermaking, are almost always found together, since papermaking requires blending and, in most cases, refining. As indicated earlier, web modifications are frequently found on the paper machine. One notable exception is pigmented coating, which can be performed at a separate site. The final group of operations, converting operations, is most likely to be separated from the papermaking operation, with the exception of tissue, cut size and folding boxboard for packaging grades. The location of the converting operation is determined by the relative costs of shipping raw materials and finished products, the speed with which products must be produced or changed and other economic factors.

This breakdown of operations also forms the basis for the organization of this book. The unit operations related to paper and paperboard manufacture are covered in each of the chapters of Section II. Section III is devoted to the major converting operations. In Section II, each of the operations is presented separately, and is accompanied by a description of the different options or treatments available and an indication of how each affects the final properties of the paper or paperboard. The reader therefore may be able to select particular segments that are relevant to certain grades of greatest interest. To make this selection, however, the reader must have some prior knowledge of which operations are needed for each grade. Chapter 3 attempts to meet that need. However, it is first necessary to define some common terms describing paper properties; Chapter 2 is devoted to that task. You are therefore encouraged to digest Chapters 2 and 3 before making excursions into the descriptive material of Sections II and III.

REFERENCES

Mikulenka, J., Mg. Ed., *Pulp & Paper 1989 North American Factbook*, Miller Freeman Publications, San Francisco, 1989.

Whitney, R. P., *The Story of Paper*, Atlanta, TAPPI, 1980.

Wilkinson, B., Ed., *PPI International Fact & Price Book 1990*, Miller Freeman Publications, Brussels, 1989.

2 Paper and Paperboard Properties

A great number of the properties and characteristics of paper and paperboard products are related to their manufacture and use. These properties are grouped and presented in Table 2.1 in the manner in which they will be treated in this book. The first category, basic properties, includes the general characteristics of the web or sheet; these can also be considered structural properties. The first four categories of properties will be discussed in this chapter; the properties pertaining to specific converting operations will be discussed in their respective chapters. Most of the tests described in this chapter are what are referred to as standard

Table 2.1. Paper and Paperboard Properties

Basic (structural) properties	**Strength properties**
Basis weight	Tensile
Caliper	Mullen (burst)
Bulk/density	Tear (internal-edge)
Moisture content and stability	Fold
Felt/wire side	**Water sensitivity**
Machine/cross-machine direction	Dimensional stability
Optical properties	Curl
Color/whiteness	Water repellency
Brightness	Loss of strength
Opacity	
Gloss	
Performance or end use simulation	
Printing—print quality	
Smoothness	
Ink receptivity or holdout	
Printing—Runnability	
Strength (tensile, pick, tear, etc.)	
Curl, wrinkle, fold (feeding problems)	
Packaging	
Barrier	
Stiffness, foldability	
Glueability, sealability	
Strength (tensile, mullen, ply bond, tear, puncture, etc.)	
General converting operations	
Uniformity (ability to be used without readjusting processing machinery frequently)	

Table 2.2. Reporting Units for Physical Testing Procedures for Paper Testing

Testing procedure[1]	Standard		Present reporting unit[2]	Suggested SI metric reporting unit	Conversion factors
	ASTM	TAPPI			
Air resistance (Gurley)[3]	D726	T460	$sec/100cm^3$	$sed/100cm^3$	
Basis weight (grammage)	D646	T410	lb/480 or 500-sheet ream	g/m^2	$\frac{\text{ream weight (lb/500 sheets}^4) \times 1{,}406}{\text{sheet area in square inches}} = g/m^2$
Bulking thickness	D527	T426	in.	mm or μm	1 point = 0.001 in. = 0.0254mm = $25\mu m$
Bursting strength	D774		$lb/in.^2$ (psi)	kPa, kgf/cm^2	1 psi = 6.895 kPa
Caliper (thickness)	D645	T411	thousandths of inch (mil)	mm or μm	1 point = 0.001 in. = 0.0254mm = $25\mu m$
Ply adhesion	D825		gf/cm width	mN/cm width	gf/cm = 9.8066 mN/cm
Standard paper conditioning	D685	T402			
Relative humidity			$50 \pm 2\%$	$50 \pm 2\%$	
Temperature			$73.4 \pm 3.6°F$	$23 \pm 2°C$	
Stiffness					
Gurley		T451	mgf	mN	1 mfg = 9.8066×10^{-3} mN
Taber			Taber unit	mN	1 Taber unit = 2.03 mN
Tear					
Edge tearing strength	D827	T470	kgf	N	1 kgf = 9.8066 N
Internal tearing resistance	D689	T414	gf/number plies	mN	1 gf = 9.8066 mN
Tensile breaking strength					
Paper and paperboard	D828	T404	lb/0.5 in. width or kgf/15mm width	newton/meter width, N/m width	kgf/15mm = 0.65378 kN/m lbf/0.5 in. = 0.53574 kgf/15mm
Tensile energy absorption		T494	$kgf \cdot m/m^2$	J/m^2	$kgf \cdot m/m^2$ = 9.8066 J/m^2

Water absorption test (Cobb)	T441	g/m^2	same or kg/m^2	
Strength factors	T220			
Beating time		min	min	
Breaking length		m	m	
Bulk		cm^3/g	same or m^3/kg	
Burst factor (index)		$\frac{gf/cm^2}{g/m^2}$	$kPa \cdot m^2/g$	$\frac{1\ gf/cm^2}{g/m^2} = 0.098066\ kPa \cdot m^2/g$
Density		g/cm^3	g/cm^3	
Tear factor (index)		$100\ gf \cdot m^2/g$	$mN \cdot m^2/g$	$\frac{100gf \cdot m^2}{g} = 9.8066 \times 10^2\ mN \cdot m^2/g$

Source: Lowe 1975, p. 92.

[1]Many physical testing units will not be changed by the conversion to metric or SI units. Thus, such tests as brightness (%), compressibility (%), Canadian Standard Freeness (ml), opacity (%), and stretch (%) will continue to be reported in the same units.

[2]Most of the suggested reporting units in SI metric terms are based on recommendations from the British Paper and Board Makers' Association.

[3]Some properties will continue to be reported in present units. Thus, such procedures as the Bendtsen air resistance and smoothness tests and the Sheffield smoothness test will continue to be reported in ml/min.

[4]If the given trade size is in 480 sheets, then factor shall be multiplied by 48/50 (or 0.960).

tests. The American Society for Testing Materials (ASTM) and the Technical Association of the Pulp and Paper Industry (TAPPI) are the two major U.S. associations that have established standard test methods for these properties. Other associations in other countries have methods that may or may not be identical to ASTM and TAPPI standards. Efforts are currently under way to attempt to establish international test methods, which will be most beneficial to the future development of the industry. A summary of testing procedures and proposed metric reporting units is given in Table 2.2.

BASIC SHEET PROPERTIES

Basis Weight

The most basic property of the sheet of paper is its *basis weight*. Paper production is measured in pounds or tons and prices are calculated per pound or ton of material. However, when the paper is used, either in communications or packaging, the user is interested in how much surface area is available for the message or to wrap around the product. Therefore, there is a need for a basic measuring parameter that is a combination of weight and surface area. Basis weight is just that. Basis weight calculations are expressions of the weight of a ream of paper of some standard size. In the USA, for example, the standard letter and writing paper size is 8½ in. by 11 in. A ream of writing paper therefore is 500 sheets of those dimensions. However, basis weight calculations are based on sheet sizes of either 11×17 or 17×22. The reason for using these larger sizes is that the mill used to produce the larger size and ship it to the converter, who would then cut the sheets to the smaller size. A ream of 20-lb (or substance 20) writing paper will weigh 20 lb only if you have 500 sheets cut to 17×22 in., and is expressed 20# (17×22-500). Unfortunately for the new person in the paper industry, different types of paper have different ream sizes. A few examples are: book papers (25×38-500); bond and writing (17×22-500); paperboard (calculated on the basis of 1000 ft^2). Even though the historical reasons for the size designations may have been lost long ago, the conventional sizes are still used in many areas.

The proposed international standard unit for basis weight is grams per square meter (g/m^2), which is called *grammage*. Use of this standard unit has not spread

Table 2.3. Comparison of ISO and North American Paper Sizes

	ISO sizes			North American sizes		
	Millimeters	Inches		Millimeters	Inches	Common use
A0	841×1,189	33⅛×46¾	NA0	889×1,143	35×45	Book
A1	594×841	23⅜×33⅛	NA1	559×864	22×34	Envelope
A2	420×594	16½×23⅜	NA2	432×559	17×22	Newspaper, mimeograph
A3	297×420	11⅔×16½	NA3	279×432	11×17	Bond
A4	210×297	8¼×11¾	NA4	216×279	8½×11	Letter
A5	148×210	5⅞× 8¼	NA5	139×216	5½× 8½	Book
A6	105×148	4⅛× 5⅞	NA6	108×139	4¼× 5½	Postcard

Source: Lowe 1975, p. 111.

too rapidly in the USA due to the persistence of older methods as well as the industry's failure to use metric sizes for cut size papers. The use of the standard term would be helped by the adoption of standard metric sizes. ISO sheet sizes are based on the principle that all necessary sizes can be cut from a 1-m^2 sheet with the dimensions of 841×1,189mm (33⅛×46¾ in.), thus eliminating the vast number of sheet sizes now stocked by paper merchants and mills. The ISO A series is based on halving the 1-m^2 sheet, designated A0, to produce an A1 sheet with dimensions of 594×841mm (23⅜×33⅛ in.). This process continues until 10 different sheet sizes are obtained. This system does not allow for any trim or scrap losses; that is, all sheets must be precisely cut. Table 2.3 compares standard ISO sizes with North American sizes, showing common uses for each size.

Caliper

Another basic consideration to the paper processor is the thickness of the paper. The thickness, called *caliper*, is usually expressed in thousandths of an inch. A sheet of paper measuring 0.0045 in. thick is called 4½ points or thousandths. The compressibility of the paper makes it necessary to use a standard measuring device or method. Some confusion can arise between caliper and *bulking thickness*, which is the average thickness when four sheets are measured together. This measurement is used primarily in the printing industry, where the printed materials will need to be folded and mailed using automatic handling machines. Bulking thickness must not be confused with bulk.

Bulk and Density

Bulk is an expression of volume per unit weight, such as cubic centimeters per gram. Bulk is the inverse of *density*, expressed in grams per cubic centimeter, pounds per cubic foot or other standard units. Calculation of these properties from the caliper and basis weight requires strict attention to units. Sometimes *apparent density* is calculated by the simple division of basis weight by caliper. *Apparent bulk* is calculated by dividing caliper by basis weight. These calculations lead to mixed units and can cause confusion. However, as long as values in matching units are compared, the practice is valid and can be helpful. Density is used by some as a predictor of paper strength since bonding in the sheet increases both strength and density.

Moisture Content and Stability

The *moisture content* of the paper is also of great importance. Paper will normally have about 5% moisture in it when dry, but that value can range from 3% to 7% depending on the type of paper and the materials used in its manufacture. In some cases shifts in moisture content within this range can cause the paper to curl, wrinkle, change dimensions or lose strength and can create other handling difficulties. Since paper is made of cellulose, which is highly hygroscopic, paper will take on water from the atmosphere or lose it to the atmosphere if the two are not in balance. The paper should therefore be made with a moisture content that will be in equilibrium with the conditions where it will be used. Since paper will come to equilibrium with the moisture in the air, the moisture content is sometimes expressed as the percent relative humidity at which the paper will be stable. The moisture content is normally measured by drying the paper to constant weight at 100°C.

Felt and Wire Side

The paper machine and its operation also build some characteristics into the sheet of paper. The first of these is the difference between the felt side and the wire side. When paper is made on the older single-wire fourdrinier (to be defined later) or on the sheet mold, it will have the imprint of the wire on one side, called the wire side. The top side, which used to be transferred to a felt when paper was made by hand, is known as the felt side. However, many modern paper machines press the paper with felts on both sides, and an increasing number of machines form the sheet between two wires, adding confusion as to which side is the felt and which the wire. The first option is to change the terms to top and bottom, which is not accurate on machines that have vertical forming sections. The difference between the two sides becomes of importance if one side is smoother than the other or if the sheet tends to curl in one direction. Unfortunately, there is no industry standard as to which side is always the smoothest or which way the sheet will curl. Some mills or machines produce paper that has marked—and even predictable—differences, but the variety of machines and processes used makes universal prediction difficult. If there are noticeable differences, the sheet is called two sided, and the printer may prefer to print one side or the other.

Machine Direction

Another characteristic property built into the sheet on the machine is its *directionality*. There is sometimes a tendency for the forming section of the paper machine to align the fibers in the machine direction as they are deposited on the wire. If this happens, it is obvious that the sheet will have a definite orientation to it. The major cause of directionality is stretching of the paper in the direction of travel as it passes through the machine. The web must be pulled from the wire and the surfaces of the press rolls or it will stick to them. When the web is being dried, it must be stretched tight to prevent wrinkling. All of this pulling of the web as it passes through the machine causes the web to be stretched in the machine direction by about 3%. Also, lack of tension in the cross-machine direction can allow the web to shrink in that direction by about 2%. The drying operation therefore builds directionality into the web. The web will expand more in the cross-machine direction when moistened and stretch more in the cross-machine direction when placed under a load in that direction. On the other hand, the web will be more resistant to stretching and expansion in the machine direction. The difference in expansion when moistened gives us the easiest way to determine the machine direction. If you moisten a small piece of paper on one side, it will curl into a tube, the axis of which will be parallel to the machine direction. However, as the paper becomes moistened all the way through, it will uncurl and perhaps even curl in a different direction.

OPTICAL PROPERTIES

Whiteness (or Color) and Brightness

When making paper for communications, we are interested in how well the paper will be able to carry the message and display it to the reader. Basic to this goal are the brightness and the whiteness of the paper. Although these two terms may be used interchangeably by the average person, they are different things to the papermaker. *Whiteness* is related to the color of the sheet and actually is the equal presence of all colors. If a sheet is truly white, it will have a high degree of reflectance and will not absorb one wavelength of light more than another. Color in an opaque or partially opaque material is determined by the amount of light energy absorbed, and the wavelengths at which the light is absorbed. For example, if white light (light that contains equal amounts of all colors) is focused on a sheet of paper and all of the incident light is reflected, the paper will appear white. If some of the wavelengths are absorbed, the color of the paper will be that of the light which was not absorbed, but was reflected back to the viewer. Blue paper appears blue in white light because it absorbs the other colors and reflects the blue.

Brightness is a special papermaker's term that is defined roughly as the amount of blue-white reflectance compared with magnesium oxide, which is considered 100% bright. The most important difference between brightness and whiteness is that brightness is measured in the blue region. The blue region was selected because the human eye and psychological system prefer blue-white to other shades. When things get old, they turn yellow, so blue is an indication of newness. Unbleached paper is yellow to brown in color and bleaching increases the blueness of the paper. Although brightness is not too different from whiteness, there is a small difference, and the person who is dealing with the paper industry must be aware of it.

Opacity

The other paper property that affects the value of paper as a communications medium is its *opacity*. To the printer, opacity means how well the paper can hide the printing on the back side of the sheet. There are different degrees to which materials will allow light to be transmitted. Opaque means that you are not able to see through at all. The paper term opacity is an indication of the degree of opaqueness, or how closely the paper approaches complete opacity. Accordingly, opacity is expressed as a percentage, with the highest opacity being 100%. An average printing paper will be around 90% opacity.

Gloss

Gloss is basically the measure of the reflectance of light from the surface of the web. The physicist uses the term *specular reflectance*, indicating that the angle of reflectance is equal to the angle of incidence. A mirror has a high specular reflectance due to its flat surface. Paper has a relatively rough surface, especially on the microscopic scale of light reflectance. The measurement is made using a focused beam of light directed toward the paper surface at an angle of 15 degrees from the surface or 75 degrees from perpendicular. The test takes this last angle as its name and is called the 75-degree gloss. There is also a 20-degree gloss, which is used for some waxed grades. The 75-degree gloss is the most common and is used for the measurement of glossy coated papers. Uncoated papers will normally be below 20% and coated papers can range from 20% to 80%.

STRENGTH PROPERTIES

Tensile Strength and Extensibility

The most basic and easily understood strength test is the *tensile strength*. Tensile is run on a strip of paper either 1 in. (2.54cm) or 15mm wide by putting it in a machine that will stretch it until it breaks. The tensile strength is defined as the force required to break the sample. Obviously, the size of the sample will affect the results and must be carefully controlled by the tester. Some pieces of tensile testing equipment allow the simultaneous measurement of the degree to which the paper is stretched before it breaks. This property is called the *extensibility* or just *stretch*. Extensibility and tensile are both very important in printing papers to ensure the ability of the paper to get through the press. They are also important for bag and other packaging papers for obvious reasons. Both are sensitive to and improved by increased bonding in the sheet.

Mullen

The puncture resistance is measured in a test called the *burst* or *mullen*. This test pushes a 1-in.-diameter rubber diaphragm against the bottom of the sheet while it is held in a special circular clamp. The test ends when a hole is popped in the paper, at which time the pressure required to pop the sheet is recorded. For obvious reasons the test is also referred to as the "pop" test. The use of the rubber diaphragm allows control of the test procedure, but is criticized for not being truly representative of what will happen to a box hit by a board, or other real world situations. The mullen test is sensitive to and positively influenced primarily by bonding, with machine direction stretch a small factor.

Tearing Resistance

Tearing resistance is readily understood as a potential problem in the use of paper, but because it is difficult to simulate, testing for this property is not easy. The first problem lies in the difference between the force required to start a tear and the force required to continue a tear that has already been started. We therefore have different tests to evaluate the two properties. However, the main test is the internal tearing resistance, or force required to continue a tear. The tester is designed to: clamp the paper to be tested, cut a notch in one side to simulate the beginning of a tear, then tear the sample through a standard distance and record the average force required to continue the tear. Tearing resistance is generally improved by the use of longer fibers and hurt by increased bonding in the sheet.

Fold

The MIT double fold test measures the number of 270-degree folds required to weaken the test strip so that it will break under a 1 kg load. The test doesn't relate directly to any end use, but does give an indication of the paper's ability to take abuse. It therefore is sometimes described as a measure of the paper's toughness.

All strength tests are sensitive to the moisture content of the paper at the time of the test, but the fold strength is the most sensitive. For this reason, TAPPI has specified standard testing conditions to be 50% relative humidity and 73°F (23°C). Paper will equilibrate with these conditions fairly quickly, but to be sure, paper should be held in these conditions at least 4 hrs prior to testing.

The performance of paper in actual printing and other converting operations will also be affected by the moisture content of the paper. Too low a moisture content makes the paper brittle, less extensible, more likely to crack when folded, susceptible to static problems and generally just less forgiving. Raising the moisture content improves these performance properties, but too much moisture can make the paper limp, perhaps too extensible and could lead to curl. A proper balance between strength tests, moisture content and extensibility is important in ensuring the paper's performance in converting operations.

WATER SENSITIVITY

Dimensional Stability and Curl

As water is absorbed by the paper web, it causes the fibers to swell and also forces them apart. The caliper of the sheet will increase slightly, and the dimensions of the sheet will also increase. As discussed with respect to machine direction, this increase will be greater in the cross-machine direction. The ability of the sheet to resist this growth is called *dimensional stability*. If the water penetrates one side of the web only, the web will *curl* away from the moistened side since that side will expand and the other will not. If the pore structure of the web is not the same on both sides, then uniform absorption of water throughout the web can still lead to curling. A nonuniform web structure is susceptible to curling from either an increase or a decrease in moisture content.

Water Repellency and Sizing Tests

Water repellency is measured by a wide variety of sizing tests. Most tests subject the surface of the web to a liquid and measure the time required for it to soak either into the web or through the web. Tests that measure only the soak-in are most sensitive to the surface sizing as applied at a size press. These tests include: (1) the water drop, (2) the contact angle and (3) the Cobb size test. The first test places a drop on the surface and measures the time required for it to soak in; the second measures the angle of contact between the drop and the paper surface. The third uses standard-sized rings to expose a controlled area of the surface to the liquid and measures the weight absorbed over a given time period. Tests that measure the rate of penetration are (1) the simple boat and (2) the Hercules sizing tester (HST). The simple boat test requires that a boat be made of the paper with the bottom 1 in. on each side. The boat is floated on a bowl of ink until an observer concludes that the ink has penetrated 50% of the bottom. This test is quite subjective and difficult to standardize. Attempts to standardize the procedure by using indicators on the surface have not been totally successful. The HST uses an electrical device similar to a brightness meter to measure the penetration of a colored ink through the web. This tester eliminates a great deal of the variability of the other float or penetration tests.

Since the penetration of water into the web causes a loss of strength, this factor has been used as an evaluator for sizing. However, this technique should not be confused with wet tensile tests, which are used to measure the effectiveness of wet strength agents. These agents do not affect water repellency, only the strength of the web when it is wet.

END USE STIMULATION TESTS

Every grade and conversion operation has associated with it a collection of tests that attempt to supply more specific performance information than can be obtained from the preceding tests. These test methods are generally referred to as *end use simulation tests* since they are designed to simulate the conditions to which the product will actually be subjected in use. These tests and properties will be considered in the chapters on converting in Section III.

REFERENCES

Casey, J. P., *Pulp & Paper*, 3rd ed., New York, Interscience, 1983, Vol. 3.

Lowe, K. E., *Metrication for the Pulp and Paper Industry*, San Francisco, Miller Freeman, 1975.

Technical Assn. of the Pulp & Paper Industry, *Official Test Methods*, *Provisional Test Methods*, and *Useful Test Methods*, Atlanta, TAPPI Press.

3 Relationship Between Operations, Properties and Grades

In this chapter we relate the grades of paper and paperboard listed in Table 1.2 of Chapter 1 to the unit operations of papermaking given in Table 3.1. There is not an easy one-on-one relationship between these two or a simple chart could be constructed and this entire book eliminated. The complications arise from the fact that more than one set of operations can be used to produce each of the grades, and one set of operations could produce more than one grade. The approach therefore will be to look at each of the manufacturing operations to see what effect it may have on the properties of the resultant paper. These properties can then be roughly connected with the properties required by the different grades of paper. It is impossible to cover all grades, but it is hoped that the exercise will help the reader see how grades that are not discussed can be analyzed and a reasonable guess made as to the ways they could be produced.

PULPING

The first set of papermaking operations, commonly called *pulping*, can also be described as consisting of selection and procurement of the raw material and treatment of that raw material to liberate the fibers from it. Included in this area is the removal of impurities from these fibers to the extent needed for subsequent operations or by the final product for which the pulp will be used.

Raw Material Selection

The choices for raw material are primarily wood fibers obtained directly from trees (*virgin fibers*) or those obtained from wastepaper (referred to as *secondary fibers*). Other fibers have been used and still are to a certain extent today. Cotton fibers give paper strength and permanence and therefore are valuable for use in money paper and paper for documents that are to be retained for long periods of time. But cotton is very expensive because of the demand for cotton fibers in clothing. If pure cotton rags can be obtained, they can be used, but the blending of synthetic fibers with cotton in clothing renders these rags practically useless to the papermaker. Straw, sugar cane and other grasses have been and are being used in certain applications because they either are more available than wood or have special properties needed in the final product.

Wood fibers can be obtained in almost as many forms as there are types of trees, since each tree will form its own distinctive, identifiable types of cells.

Table 3.1. Unit Operations of the Paper and Paperboard Industry

Pulping	Stock preparation	Papermaking	Web modification	Converting
• Selection and preparation of raw materials –Selection –Procurement –Preparation • Liberation of fibers –Mechanical –High-yield –Chemical –Secondary fibers –Bleaching	• Fiber modification theory –Fiber structure –Bonding theory –Fibrillation theory • Fiber modification equipment –Beaters –Refiners • The furnish –Metering –Blending –Cleaning –Screening –Nonfibrous additives	• Preforming considerations –Approach flow –Headboxes –Materials balance • Forming devices –Fourdrinier –Cylinders –Multiwire machines • Consolidation of the web –Presses –Press fabrics –Pressing theory • Drying –Theory –Methods	• Surface modification –Sizing –Calendering –Pigmented coating • Physical modifications –Rewinding –Cutting	• Printing –Letterpress –Flexography –Gravure –Offset-litho • Corrugating –Corrugating –Boxmaking • Packaging –Flexible packaging –Folding boxboard –Rigid boxes –Cans and tubes –Labels • Cut size and office supply –Typing/writing –Copy papers –Business forms –Computer • Tissue and related –Tissue –Toweling –Napkins

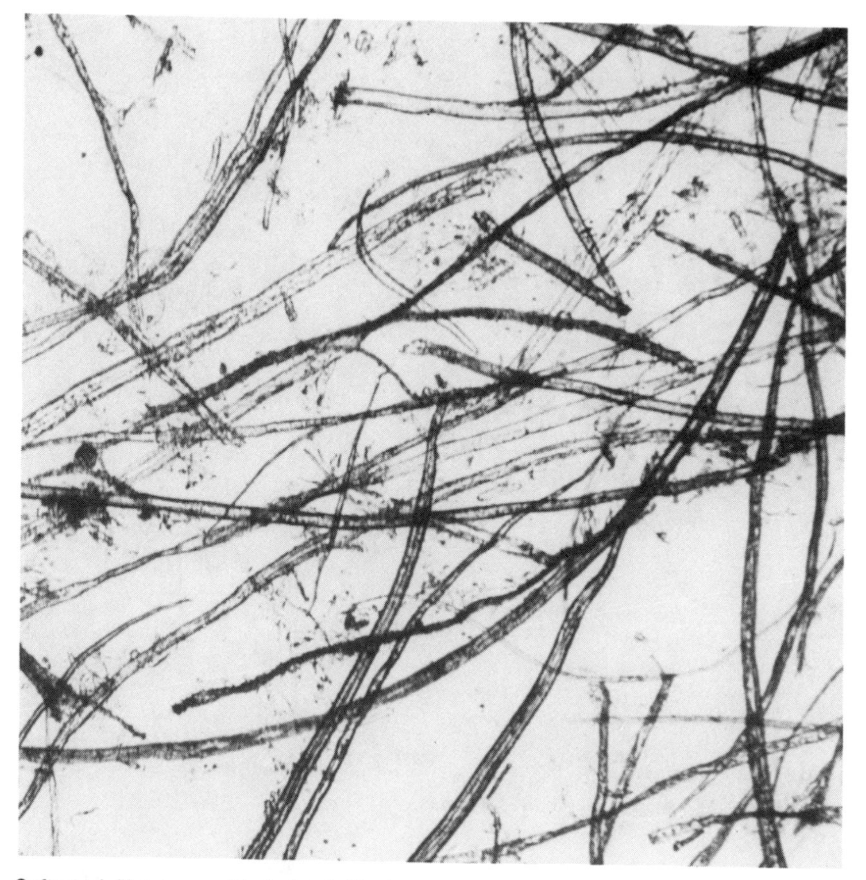

Softwood fibers magnified about 50 times.

However, the differences between some species are too small to affect the paper properties. Therefore, we group wood fibers into two major categories: the hardwoods and the softwoods. Hardwoods are the broad-leafed trees, such as maple, oak, birch, gum, and so forth. Softwoods are the evergreens or needle-bearing trees, such as pine, spruce, fir, and others. Softwoods generally have longer fibers, which will contribute to greater strength in the paper. These same long fibers can make the paper rough or coarse feeling, however. The hardwoods give us fibers that help to fill in the sheet of paper, making the sheet smoother, more opaque and usually better for printing. Softwoods are frequently the only fiber used in grades where strength is needed and the coarseness can be tolerated. Hardwoods, on the other hand, are seldom used alone due to the low strength of the paper produced. Most paper and paperboard is made from a blend of both types balanced to achieve the desired final properties.

The choice between virgin fibers and secondary fibers is made on two major points: strength and purity. Secondary fibers are generally lower in strength than

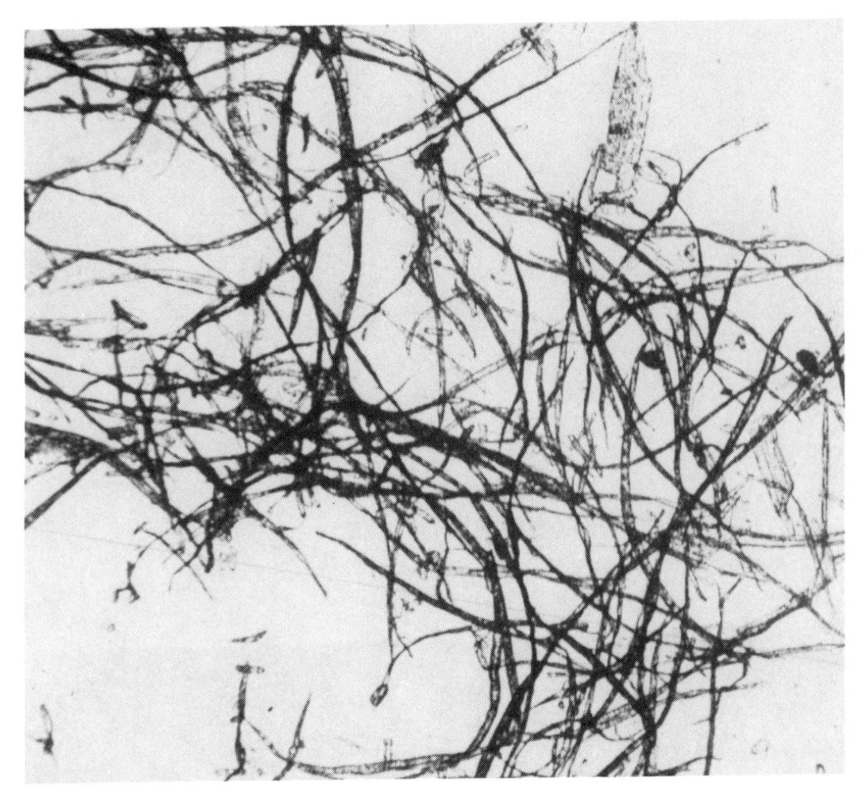

Hardwood fibers magnified about 50 times.

if the same blend of fibers was obtained as unused pulp. Secondary fibers can be cleaned and bleached to produce high-quality papers, but the cost of these added operations may negate the savings sought by using the lower cost wastepaper as a raw material. Some secondary fibers are being cleaned and used in tissue or as a portion of the furnish, or formula, for printing grades. The major application of secondary fibers is in products where cleaning is not needed. Combination boxboard, more commonly recognized as cereal cartons or other boxes made from board that has a gray layer in the center, is the largest use of secondary fibers. These grades are known as combination furnish–bending and nonbending and are listed in Table 1.2. The secondary fiber, or "chipstock," made primarily from old newspapers and corrugated containers is used in the center of these grades. In this application it gives bulk and strength to the paperboard, but can be covered with white pulp and coating to give the desired printing characteristics.

Liberation of Fibers From Wood

Once the raw material has been selected, we need to liberate the fibers. *Mechanical* methods for pulping or liberating fibers produce slightly yellow paper of fair or low strength commonly used in manufacturing newsprint and other publication grades. These processes are used primarily on softwoods to maximize the strength of the pulp. Mechanical pulps are not likely to be the sole pulp used, and will be blended with stronger, whiter grades to improve the quality and permanence of the paper produced. *Chemical* methods for liberation use either softwoods or hardwoods, or a blend of the two. The kraft pulping process, a chemical pulping operation, will generally produce a stronger paper and is more common than the sulfite process. Strength differences are not the only reason for the preference for kraft pulping. Waste products from kraft pulping operations can be reused more readily, and other ecological considerations also favor the kraft process over sulfite. Unbleached softwood kraft is found in packaging applications such as sacks, bags, tubes, cans, drums and some folding cartons where maximum strength and water resistance are needed. Bleached softwood kraft is used in almost all grades of paper because of its whiteness and strength. It is used alone in some packaging grades but is generally blended with bleached hardwood kraft for improved smoothness, opacity and printability. Sulfite pulps are generally whiter than kraft in the unbleached form and therefore have been used up to about 20% in the unbleached form to add strength to newsprint. Bleached sulfite has been used in tissue because the fibers can make a softer paper than kraft, and in printing or bond papers because the fibers are purer and therefore give greater permanence to the paper. However, ecological problems have almost eliminated all use of sulfite pulping. Most grades of paper and paperboard are made from a blend of different pulp types to give the final product the optimum combination of properties at the least possible cost.

Between mechanical and chemical processes are the *high-yield* processes, which use some chemical treatment and some mechanical treatment to liberate the fibers. As such, they have some of the advantages and disadvantages of each process—producing pulp that is not as strong as kraft or as bright as groundwood. High-yield pulps give a degree of stiffness and strength to paper that makes them ideally suited for use in corrugated containerboard manufacture. They can also be employed in some of the other applications where unbleached kraft has been indicated. The name "high yield" indicates another difference between the pulping operations. Full chemical pulping operations dissolve the natural glues (lignin) in the tree to liberate the fibers. The chemical reactions also remove some of the cellulose, leaving us with a lower yield from the pulping operation. Full chemical pulps may have a final yield of only 50% of the initial weight of the wood, whereas the mechanical pulps can yield more than 90%. Yields from the combination processes fall somewhere in between, depending on the extent of chemical treatment.

Another operation generally associated with pulping that reduces the yield of the pulping operations is *bleaching*. Bleaching can also be a purification operation since the chemicals used are designed to react with and remove colored materials from the fibers. The colored materials are largely from the natural glues in the wood. Bleaching will have little effect on the strength of the resultant paper unless the pulp is bleached extensively or to very high brightness, which can reduce strength a little. The major reason for bleaching is its effect on the whiteness or brightness of the paper.

STOCK PREPARATION

Within the second set of papermaking operations, commonly known as *stock prep* or *beater room operations*, we find three categories. The first, fiber modification or refining, is performed on most grades of paper but may not be needed to any great extent on groundwood or secondary fibers. There are basically two results to be obtained from refining, either cutting of the fibers or fibrillation. Cutting will shorten the fibers, possibly reducing the strength of the paper but at the same time improving the uniformity, or what is known as the formation. Cutting has less effect on strength of the paper than fibrillation. Fibrillation will be shown in later chapters to be the physical modification of the fibers to facilitate bonding between fibers. As such, it is the process most directly related to the development of strength in the paper, after the initial selection of the raw material and the pulping method. The degree of refining to which the fibers are subjected can reshuffle some of the strength relationships mentioned in the previous discussion to a certain extent, but not completely. Refining will usually be used as a "fine-tuning" type of operation since refining will not allow us to bring a mechanical pulp up to the strength of a full chemical pulp.

Since most papers are made from blends of fibers, chemicals and pigments, we need to have operations that control the ratios and ensure the proper mixing of the ingredients. The actual operations used will have less effect on some paper properties than the selection of the additional chemicals and pigments. The chemicals generally give water resistance and the pigments contribute to the brightness and opacity of the paper.

PAPERMAKING

Within the third set of operations the second group, forming devices, gives the products their major physical differences. The breakdown used in Table 3.1 is based on manufacturing machinery classifications and not on the grade structure; therefore Tables 1.2 and 3.1 do not fit together well. Single-layer paper grades are primarily any lightweight, thin or flexible paper or paperboard. The oldest and most common machine for making lightweight paper is the fourdrinier, but as we have increased speeds and production rates and made the machines more specific for certain grades, we find that there are specialized machines for tissue, newsprint and the publication grades. Heavier grades of paper and paperboard are usually produced by combining several layers of fibers or webs of paper. The combining is done immediately after forming, while the webs are still wet, so that the hydrogen bonds will join them together in the same way as the fibers that are deposited in the individual webs bond together. Combination boxboard, which has the less expensive gray wastepaper in the center and the more expensive white layers on the outside, is an excellent example of the value of combining webs. It would be impossible to form the entire web at one time and still keep the layers separated as they are in this product. Furthermore, we will see that the speed of formation of the web is reduced by the thickness of the web, making the formation of thick webs a slow process and one that is not competitive with the faster forming and combining techniques. Even in the case of the solid bleached board used for milk cartons or ice cream boxes, the thickness is often obtained by combining identical layers during the forming operation.

The third operation listed in this area, consolidation of the web, is again used on all grades of paper and paperboard. The web is deposited on a forming wire or screening device at a thickness greater than the final thickness of the product being produced. The web or webs (after combining) are pressed between rollers to reduce the thickness, bring the fibers closer together to promote bonding and remove water.

All grades of paper and board will be subjected to varying amounts of pressing, and the differences in the final product are not too easily traced to variations in pressing. Roughly speaking, bulky products like tissue will receive the least pressing, and in some machines will not be pressed at all. The stiffest, most dense paper or paperboard, such as glassine and some construction boards, will be pressed the most.

All paper and paperboard must also be dried, since water is used to form almost all paper and must be removed. There is little variation in the manufacturing methods used for drying and little effect on final properties, except for the yankee dryers used for tissue and machine-blazed papers; these allow the creping of tissue or the formation of the smooth, flat surface of the glazed paper.

WEB MODIFICATION

Web modification operations are used in every grade listed in Table 1.2. The most common operation is the use of the calender stack at the end of the paper machine. The calender stack is simply a stack of steel rolls through which the paper is passed. The rolls smooth the surface of the web and may reduce the caliper, or thickness.

The second most common treatment, but one that physically occurs before calendering, is surface sizing. The surface size is usually applied in the paper machine after the web has been formed, pressed and nearly dried. The sizing materials are usually starch solutions, which are applied as the web is passed between two rollers. The sizing agent is intended to smooth the surface; it also increases the resistance of the surface to water and to being disrupted or pulled apart by the forces created in the printing press when the ink is split between the paper and the application surface. Obviously, this treatment is most important for printing and writing grades. Packaging grades may use this treatment, but will frequently need more resistance than can be obtained with a size press, so they will be coated with other materials in what is classified as a converting operation (listed as a separate group of operations in Table 3.1). Tissue grades, which do not need water resistance, will receive mechanical surface treatment in the form of creping on the paper machine or embossing to make the surface softer to the touch.

Since paper or paperboard is produced on a machine that is normally in the neighborhood of 20 or more ft wide and the user of the product requires smaller rolls or even sheets of the product, there is a need to modify the physical shape of the web to suit the consumer's needs by rewinding or sheeting.

For the operations not covered and grades not discussed, it should be possible for the reader to analyze the properties needed and develop potential strategies for manufacturing them. A summary of the general relationships between properties and operations is presented in Table 3.2. A plus (+) indicates that the operation on the left can positively influence or increase the property at the top of the column; a minus (−) indicates a negative influence or decrease in the property. This type of generalization is dangerous, however, since results can differ in special cases or one operation can negate another when they are performed in combination. In short, the table is intended as a summary of average conditions and cannot be considered foolproof or absolute.

Table 3.2. Relationships Between Operations and Properties

	Property												
Operation	Basis weight	Caliper (thickness)	Bulk	Density	Strength	Stiffness	Color[1]	Permanence	Opacity	Water resistance	Printability	Smoothness	Softness
Pulping													
Raw material													
Softwood				+	+	+			−	+			
Hardwood			+		−				+		+	+	
Secondary			+		−		G		+		+		
Pulping method													
Mechanical			+		−		Y	−	+		+		
Chemical					+		B	+		+			
High-yield				+		+	B			+			
Bleaching			+		−	−	W	+				+	+
Stock preparation													
Refining				+	+	+			−				
Nonfibrous													
Chemicals						+		−		+	+		
Pigments			+		−		W		+		+	+	+
Papermaking													
Forming device													
Fourdrinier	−	−											
Cylinder													
Multiformer	+	+	+			+							
Pressing				+	+							+	
Drying													
Web modification													
Sizing					+	+				+	+	+	
Calendering				+							+	+	
Pigmented coating							W		+		+	+	

[1] G=gray Y=yellow B=brown W=white

Section II
Unit Operations of Papermaking

In this section the individual operations needed to produce all paper and paperboard products are presented and discussed. If the reader has been able to determine which of the operations are most important to the grades or products he is dealing with, he may be able to proceed directly to the chapters discussing those operations. However, the interdependence of all the operations and the importance of each in determining the properties of the final product cannot be overemphasized and the reader is encouraged to study all of the chapters. The chapters are presented in the order in which the fibers are processed into paper and paperboard and subsequently into salable products:

4 Pulping

Paper is a thin or layered network of randomly oriented fibers bonded together by their ability to form hydrogen bonds. This basic definition of paper can be applied to any paper or paperboard product. We would have to add a long list of qualifying statements to completely define any one paper or paperboard product, but all products would still have these things in common: (1) fibers that are able to bond to one another and (2) fibers that are formed into a layered structure with random orientation. This definition also places prerequisites on the raw materials that can be used and on the unit operations needed to prepare the fibers and to form them into paper.

FIBER SOURCES AND THEIR PROPERTIES

The first prerequisite for a papermaking fiber is that it be able to bond to other fibers without the addition of glue or adhesive to the structure. We will see that *cellulose fibers* have this property and therefore are the prime raw material for papermaking. Fortunately, this same ability to bond through a hydrogen bond helps make the second part of the definition possible. Cellulose fibers are readily dispersed or suspended in water, which serves as a carrier to deposit them in the layered network of random orientation. The formation of this network structure is the subject of Chapter 6. In this chapter we are concerned with the nature of the cellulose fibers—where they are found and how they are liberated from raw materials and processed into a suspension of individual fibers.

Cellulose fibers found in most living plants can be used for papermaking. These fibers can be separated and dispersed in water, and can therefore be deposited from the water suspension in a random network. The polarity of water and the presence of hydroxyl groups on the fibers conspire to satisfy the need for the fibers to bond to one another, through the hydrogen bond. By selection of the proper source for our cellulose fibers we can obtain the strength or smoothness of surface needed for the different applications to which the paper will be put. Of course, one prime consideration must be the ready availability of the desired raw material. Many plants can supply fibers that can be used to make paper. However, at the present time wood is the predominant source for fibers. As the world-market demand for paper expands and as the demand for wood grows in other applications, alternate fiber sources may become more important. Regardless of the source, however, the requirements will be the same and we must decide what the most important demands are to select the raw material that can satisfy them.

Since wood is the predominant raw material at the present time, the major emphasis of this book will be on the use of wood. Wood is composed of cellulosic cells, as is most living plant tissue. However, many different forms of cells are present in different plants. Trees grow and develop through the division of special cells under the bark, known as *cambial cells*, that produce both the bark and woody tissue. In the spring when growth is rapid, the woody cells swell and become rather large. During the summer or slower growth periods, the cells swell only slightly. These differences in expansion following cell division give the tree its characteristic annual growth rings. The ratio of spring wood to summer wood will have an important influence on the properties of the fibers and the paper produced from the wood.

Immediately after the cells are formed, they are filled with a living material that deposits more cellulose on the inside walls of the cell, developing the final fiber wall thickness. When the cell has grown to its full size, the living material dies, leaving a hollow cell, or *fiber*. The fiber is essentially a hollow tube with holes or pits in its walls connnecting it to other fibers. These fibers are used by the tree to conduct fluids up to the leaves, where photosynthesis takes place, and to carry sugars back to the growing regions of the tree. After conducting fluids for several years, the fibers are no longer needed for conduction since new cells are becoming available; the older cells become plugged and may dry out. These nonconducting cells near the center of the tree are called *heartwood;* the younger cells that are still conducting are termed *sapwood.* Trees with a large amount of heartwood are not as desirable for papermaking since heartwood fibers will tend to be stiffer, not bond as well or form as smooth a sheet as sapwood fibers.

Beside conducting liquids, the fibers must also support the tree and store liquids to maintain life during dry periods. All of these demands cause the tree to produce different forms of fibers. Furthermore, the two major categories of trees, *hardwoods* and *softwoods*, contain different types of fibers or cells. These differences in fiber types help explain why paper made from different types of trees will have different properties. For the papermaker the difference between these two categories is, as noted earlier, that the softwoods have longer fibers, as indicated in Figure 4.1, and will therefore facilitate the production of stronger paper.

Figure 4.1. Relative sizes of hardwood and softwood fibers

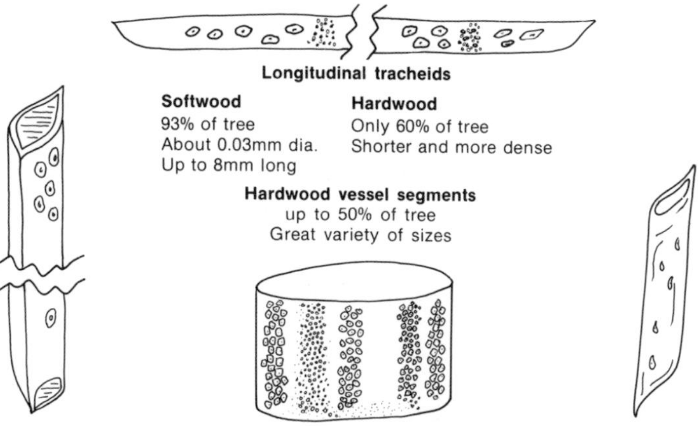

Longitudinal tracheids
Softwood
93% of tree
About 0.03mm dia.
Up to 8mm long
Hardwood
Only 60% of tree
Shorter and more dense
Hardwood vessel segments
up to 50% of tree
Great variety of sizes

However, the longer fibers of the softwoods may also be larger in diameter and thereby produce a paper that will be coarser or rougher on the surface. The hardwoods, having a greater percentage of smaller fibers, will tend to fill in the sheet more, producing a smoother surface and possibly a more opaque sheet. Although the two types of trees may be grown, harvested and pulped separately, they are usually combined to some degree during the papermaking operation so that the final sheet of paper will possess some of the desirable characteristics of each type.

In addition to the differences in the physical shapes of the fibers found in hardwoods and softwoods, there are also chemical differences between the two. Photosynthesis in trees produces primarily glucose, which is converted into cellulose, but may also produce other sugars that are not included in the cellulose structures found in the fiber wall (described in greater detail in Chapter 5). These other sugars are called *hemicelluloses* and are found in different degrees in the two types of trees, as indicated in Table 4.1. The other component listed in Table 4.1, *lignin*, is the phenolic glue or cementing substance created by the tree to hold the fibers together. These differences in chemical composition are less important in influencing paper properties than are the physical characteristics.

Beside the gross differences between hardwoods and softwoods, there are more subtle differences between species, and even within species, depending on where the tree is grown, its age at harvest and other factors. Most of these differences are beyond the scope of this text, but should be recognized as possible complications or reasons why absolute statements linking paper properties to fiber properties cannot be made. We do need to consider the importance of geographic location of the trees. If the trees are grown in a region where there is abundant rainfall and a long growing season, the fibers will tend to be larger and the paper produced will tend to be coarser. These differences are readily seen in the fairly coarse southern pines used for wrapping and bag papers and the finer-fibered northeastern spruces and pines that have been favored for writing and printing grades. As the paper industry expands into new world markets and new raw material sources are tapped, we often find that species which are not useful in one region will be acceptable in another, due to differences in climatic conditions. Furthermore, we may find that nonwood fiber sources may become more important in our future. Other fiber sources are already recognized as completely acceptable as raw materials for papermaking, and quite a bit is already known about the properties to be expected from their use. If and when the economics are right, other sources will be utilized.

Wastepaper, also known as *secondary fiber,* is becoming increasingly important, both in an effort to save trees for other uses and to remove some of the

Table 4.1. Chemical Composition of Typical Softwoods and Hardwoods

	Softwoods	Hardwoods
Cellulose	50%	50%
Hemicelluloses	20%	30%
Lignin	30%	20%

Table 4.2. U.S. Consumption of Fibrous Materials in Pulp, Paper and Board Manufacture

	1988	1987	1986	1985	1984	1982	1980	1970
	(000 short tons)							
Woodpulp, total[1]	**62,647**	**61,231**	**59,882**	**56,639**	**50,309**	**50,186**	**51,804**	**43,192**
Wastepaper, total	**17,181**	**16,618**	**15,419**	**14,818**	**14,944**	**13,563**	**12,583**	**10,594**
Mixed	2,422	2,169	2,115	2,081	2,136	1,980	2,686	3,393
News	3,185	3,143	3,140	3,031	3,018	2,790	2,116	3,140
Corrugated	8,665	8,345	7,502	7,234	7,231	6,486	5,628	3,511
High-grade pulp substitutes	1,323	1,433	1,253	1,111	1,248	1,237	1,302	1,377
High-grade deinking	1,585	1,529	1,409	1,360	1,310	1,148	851	493
Other	**275**	**273**	**288**	**301**	**346**	**331**	**556**	**780**
Cotton linter pulp	**146**	**105**	**111**	**95**	**92**	**65**	**49**	**46**

1. Woodpulp is also consumed in many nonpaper and board products, such as chemical cellulose, molded pulp products, etc.

Source: Bureau of the Census as given in *Pulp & Paper 1989 North American Factbook*, p. 264.

fibrous waste material from the waste stream and reduce its deposit in sanitary landfill sites, which are becoming scarce. The increase in use of wastepaper in the USA can be seen in Table 4.2. The fibers obtained from wastepaper are predominantly wood fiber. However, the fibers will be mixed and not all of one species or type, and, as we will see in later chapters, chemicals and minerals will have been added to the paper that may complicate its reuse. Even if the wastepaper to be reused were all of one type and had minimal chemical and mineral additives, the fibers still would not be the same as virgin fibers. We will see in later chapters that the act of preparing the fibers for papermaking changes or modifies them physically. Furthermore, the act of papermaking causes the fibers to bond together. In order to reuse the fibers from wastepaper, these bonds must be broken to separate the fibers once again. It is practically impossible to break the fibers apart without damaging them somewhat. All of these factors conspire to make paper produced from secondary fiber sources different from paper made from virgin fiber. The differences are hard to generalize because of the influence of the chemicals and treatments that may be found in the wastepaper, but generally paper made from secondary fibers will not have the same potential for strength.

WOOD PROCUREMENT AND PREPARATION FOR PULPING

Harvesting

Since wood is the major raw material for papermaking, it is appropriate that the methods for procuring and preparing wood be considered briefly. In order to be used in the mill, the wood must obviously be harvested, or *felled*, and transported from the forest to the mill. Pulpmill operations are such that they must operate 24 hr/day without interruption. When the mill is shut down and restarted, considerable waste of raw material occurs. It is therefore necessary for the mill to maintain a storage facility to ensure continued operation even though the wood supply might be interrupted. In some regions of the world where the winters are severe, this means that the mill must be able to stockpile several months worth of wood.

Methods of harvesting trees are varied and are determined by the location of the trees and, of course, the economics of the operation. The simplest operations are those in which the trees are felled by lumberjacks traveling through the forest on foot, carrying axes, saws or power chainsaws. The felled trees then must be removed from the forest, usually being dragged by a powered vehicle called a *skidder* to a central collection area where they are transferred to trucks or railcars. The skidder vehicle will be designed or selected to suit the terrain and may travel on treads or may even use large, air-filled tires to travel through wet or swampy land. The movement of the logs through the forest is facilitated by removal of the branches; and since they are of limited utility to the lumber mill or pulpmill, they are usually removed in the forest. If the wood is to be used for lumber, it is desirable to leave the tree whole—that is to say, the main trunk of the tree will not be cut into shorter sections. However, cutting the tree into shorter sections makes it easier to handle both in the forest and on highways or railroads. Helicopters and lighter-than-air aircraft have also been used in a few cases to lift logs from extremely inaccessible terrain; however, their use is very expensive and the wood must be very valuable to justify such measures. Rivers have been used to transport logs to the mill almost as long as there have been mills. The mill requires water for its operation and therefore is usually located on or near a river, allowing the logs to be floated down to the mill and the pulp or paper to be shipped away from the mill. River transportation also keeps the logs wet, which is desirable for the pulping operation. Traditionally, logs have been cut during the winter, loaded onto the frozen surfaces of streams and rivers and carried down to the mill with the spring thaw and runoff. The tumbling and rubbing action of the logs against each other can help in the removal of some of the bark. Concern for environmental protection has reduced the prevalence of floating logs to the mill and has had an impact on all of the other methods of harvesting.

Where the terrain permits, vehicles equipped with a scissors-like attachment can move through the forest and cut each tree with one "bite." The cutting blades of these machines are hydraulically loaded and can be sized to handle trees up to nearly 1m (3 ft) in diameter. These felling machines may also be equipped to hold the tree while it is being cut and then lay it on the ground, preventing damage that might be caused by allowing it to fall. Smaller trees may be picked up by some felling machines, fed through a series of knives that remove the branches and perhaps even cut into bolts, or short sections, of the desired length. The machines may be designed to collect the bolts and stack them in piles to be skidded to the road or loading areas. One such machine has been designed to cut the tree at the ground, pick it up, feed it through a "delimber," cut it into bolts and collect the bolts and load them onto a wagon or deposit them on the

Roundwood storage on either side of the water flume that carries the logs to the barker and chipper shown at the rear of the photo.

ground. If the trees are planted in straight rows and the land is level, this machine can harvest and pile up to 5 trees/min at top speed, or average 3 trees/min over a day's time.

There are also machines available that can remove the bark and chip the trees in the forest, allowing chips, rather than logs, to be hauled to the mill. The methods that can be used to harvest and haul trees to the mill are obviously diverse. In the final selection of a method, each mill needs to consider all of these (and many other) factors: geography, climate, environmental impact, species of tree, whether lumber will be manufactured and labor and equipment costs and availability.

The logs that are harvested in the forest need to be hauled to the mill and stored prior to their use. Roundwood, as logs are called, may be stored in a variety of different ways. The logs may be stacked in nice, neat piles, which although easy to handle are expensive to develop and maintain. Modern equipment consisting of large cranes with grapple claws are able to lift random stacks of logs from piles, and this type of random stacking or piling is more prevalent in today's mills. Water is also frequently used as a transport medium, especially in the final transfer stage when the logs are removed from their storage area in the woodyard and moved into the mill area. In such an operation, the logs would be dumped into a water flume, which carries the logs into a chain-conveyor ladder called a *jackladder* that carries the logs up the chute and delivers the logs to the barker or chipper.

Bark Removal

Pieces of equipment designed to remove bark from the logs are called *barkers*. Barkers come in two basic varieties—hydraulic and mechanical. Hydraulic barkers, as the name implies, require the use of water to remove the bark from the trees. The water is pumped by high-pressure pumps into jet squirters, which impinge on the surface of the tree or log and blast off the bark. The hydraulic barker does a very

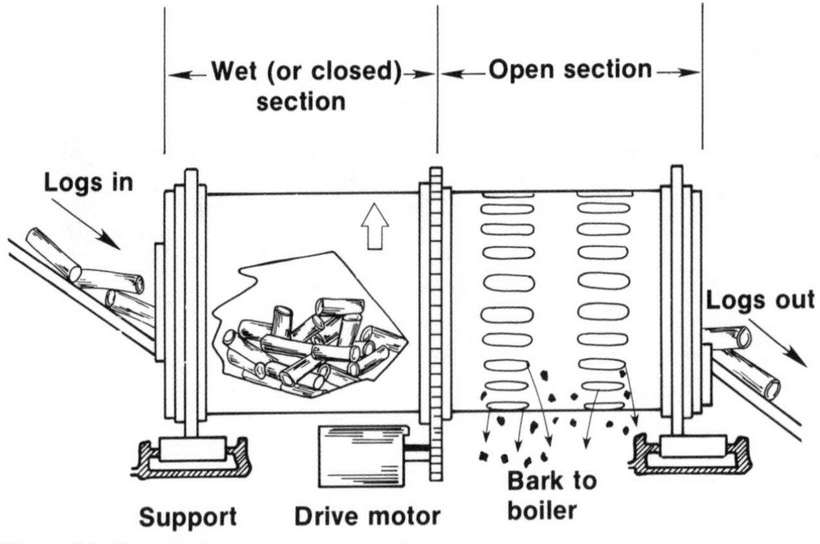

Figure 4.2. Drum barker

efficient job with very little damage to the wood and is most suited to the barking of logs that are large in diameter or very long. It has therefore found application in operations where a lumber mill and a pulpmill are operating together, or in other applications where treelength logs need to be handled.

Mechanical barkers can be broken down into major categories: those that handle a large number of logs at the same time, and those that treat individual logs. The type that treats a large number of logs at a time is called a *drum barker*. A representative drawing of a drum barker is shown in Figure 4.2. The drum barker is a large cylinder open at both ends, or perhaps partially closed at the exit end to control the rate of travel of the logs through the barker. Logs with bark on their surfaces are loaded into the higher end of the drum. The drum is rotated, causing the logs to tumble over one another creating a rubbing action that strips the bark from the logs. The initial part of the drum barker may be closed to retain water and help soften the bark and remove it from the trees. Later sections, or the entire drum barker, may be open, to allow the bark that is removed from the logs to fall out through the openings and be transported away. In some applications the drum barker will be partially submerged in water, to keep the logs moist and to carry the bark away from the barker. In order to handle the logs, the dimensions of the drum barker need to be fairly large. Drums are generally 12 to 15 ft in diameter and a barker is usually made up of sections perhaps 15 ft long. Since a single drum barker may have up to five sections, each barker represents a rather substantial piece of equipment. The size of the logs to be handled in a drum barker needs to be controlled; bolts will generally be 4 to 6 ft long. The diameter of the wood also must be kept above a certain minimum so it will not fall or stick out of the bark removal holes. Drum barkers do a good job of removing bark from large amounts of wood in a fairly efficient manner. However, the

Discharge end of a pair of drum barkers showing the log conveyor in the foreground carrying barked logs to the chipper.

tumbling action in the barker can cause some damage to the ends of the logs, producing a fraying or brooming effect.

Figure 4.3 shows one type of individual-log mechanical barker, the *ring knife barker*. There are a number of different varieties of such barkers. As shown in Figure 4.3, the wood is carried through a ring of knives that rotates around the log, stripping bark from its surface. The ring knife barker can also remove good

Figure 4.3. Ring knife barker

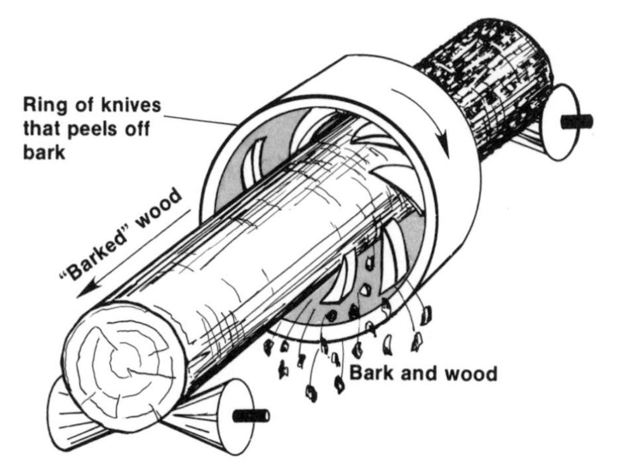

fibers with the bark if the wood is at all soft. Therefore, variations of this main idea have been developed that are more gentle to the wood. Variations include using chains, which will tangle in the bark and tear the bark off the surface of the log, or utilizing rough-surfaced wheels that rub on the surface of the log. Some mechanical barkers will also cause the log to rotate as it passes through, to increase the rubbing action and aid in removal of the bark. In any case, all mechanical barkers are subject to two problems: removing good wood along with the bark if the wood is soft and failure to remove all of the bark if the bark is too hard or too securely held. For these reasons, barking operations function better if the wood is wet and green.

Chipping

The first step in the manufacture of chemical pulp is the *chipping* operation; the logs need to be reduced to small chips to allow the chemicals, or *cooking liquor*, to penetrate the fibers and dissolve the lignin. The route by which the chemicals penetrate the chips has an important ramification for the papermaker: the pulping operation is dependent on the nature of the chips. The penetration of the liquor into the chips has been shown to take the same route as the fluids that the tree conducts—namely, through the center of the fibers. The liquor penetrates into the center of the fibers at the end of the chip, then passes from fiber to fiber, working its way through the fiber walls to the lignin between the fibers. This mechanism of reaction seems to be occurring in the most undesirable direction, since most lignin is found between the fibers. However, when one realizes that the areas between the fibers where the lignin concentration is the highest are completely blocked from penetration, and furthermore that the fibers functioned in the tree as conduits for fluids and are therefore designed to allow liquids to flow through them, it becomes more plausible that the liquor should penetrate by this route. Thus it can be seen that chips must be of a fairly uniform size and must be screened to remove small and large chips.

The chipper is shown in Figure 4.4 to be a large flywheel with knives mounted in its surface. The logs are fed to the flywheel through a chute at an angle such

Figure 4.4. Chipper and chipper detail

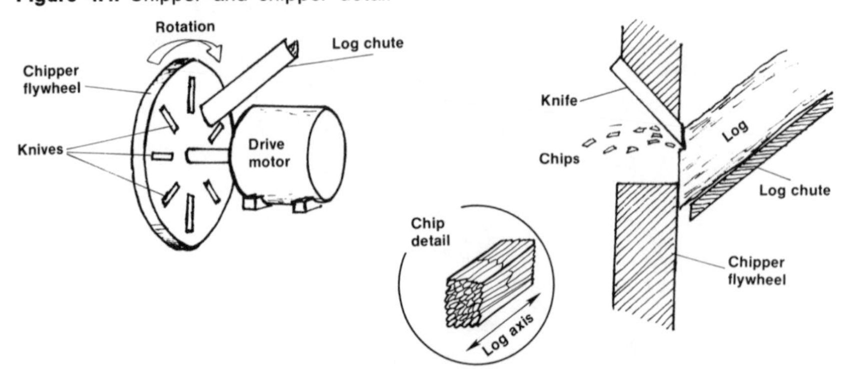

View of logs going down the chute into the chipper.

that the entrance of the knife into the log will be at a sharp angle to the axis of the log. The drawing in the lower part of Figure 4.4 shows a detail of the chipping operation. The knife enters the log at a sharp angle, cutting and splitting the log at the same time. The chip in the lower left corner shows that the longest dimension of the chip is in the direction parallel to the centerline, or axis, of the log. The length of the chip can be controlled by a combination of the speed of the flywheel, the angle of the log and knife, and the extension of the knife beyond the face of the wheel. The other dimensions of the chip are less easily controlled, being dependent on the same parameters and also on the moisture content of the log and its splitting or splintering tendency. However, since the penetration of the liquor into the chip proceeds mainly through the fibers, the length of the chip is the most important dimension.

Unscreened wood chips.

GROUNDWOOD PULP MANUFACTURE

The basic *groundwood pulp process* is deceptively simple; barked logs are held against an abrasive stone, which tears the fibers from the log, and water is used to wash the fibers from the stone. In fact, the earliest groundwood operations were that simple. Naturally occurring sandstone was shaped into grinding stones about 3 ft in diameter, and several stones were assembled side by side on a shaft to give a grinding surface up to 4 ft wide. Such early stones were driven by waterwheels at relatively low speeds. Modern stones are not solid, but rather are covered with blocks of synthetically produced abrasive materials of controlled grit size. Such synthetic stones can be made more than 5 ft (1.5m) in diameter and up to 6 ft (2m) across their face and can be motor driven at speeds up to 10,000 ft/min (50m/sec). Modern stones are also given a special surface treatment or burr pattern to help control the removal of the fibers and their quality. Figure 4.5 shows a representation of the cross section of this pattern and a log being ground. The stone is patterned with a series of grooves that cause the surface of the log next to the stone to be alternately compressed and relaxed. It is also theorized that the groove pattern pumps water in and out of the wood fibers, aiding in their removal and also helping to control the temperature of the wood. The alternate compression and contraction generates heat in the wood, which softens the lignin and fibers and aids in their removal. However, if too much heat is generated, the wood will char and the fibers will be destroyed. The pattern on the surface of the stone must be carefully maintained by frequent application of a small cutting wheel or burr to the surface of the stone.

Figure 4.5. Basics of grinder operation

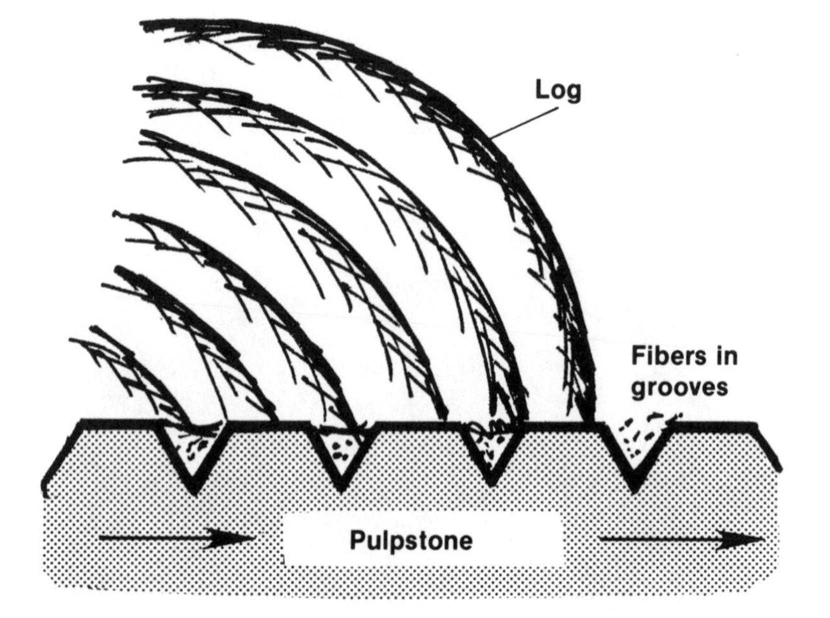

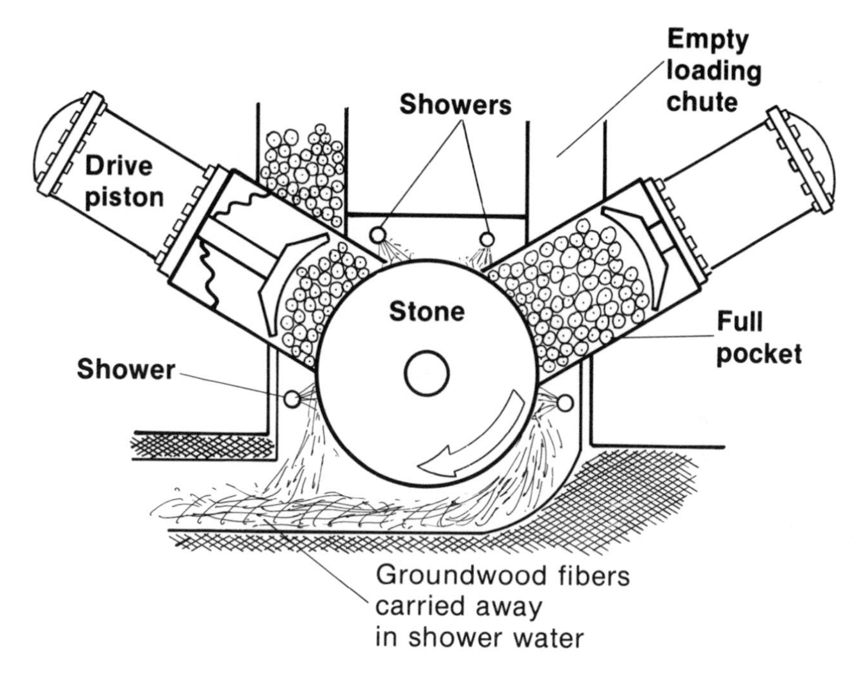

Figure 4.6. Two-pocket grinder

Figure 4.6 shows a common form of grinding machine, the *two-pocket grinder.* The machine has two pistons, or pockets, which are used to push the logs against the surface of the revolving stone. It should be noted that the logs are pressed against the stone such that their axes, or longitudinal centerlines, are parallel to the center of rotation of the stone. This method of grinding is most efficient in removing whole fibers from the wood and also in grinding the whole log and reducing waste from small pieces or slivers of wood that are too small to be held against the stone and that might otherwise pass on through with the fibers. The use of two or more pockets allows one pocket to be filled while the other is still at least partially full and being pressed against the stone. Maintaining a constant load on the stone in this manner is important in the maintenance of both constant speed and temperature of the stone. Other feeding-arrangement designs that serve the same function are available.

The groundwood pulp from the grinders needs to be screened to remove large fiber bundles known as *shives* and other large materials such as slivers or underground pieces of wood and knots. After screening, the groundwood pulp is ready for any subsequent treatment or may be sent directly to the paper machine. If unbleached pulp is desired (for newsprint, for example), the pulp will not require any further treatment. It is possible to bleach the groundwood pulp to improve the whiteness and permanence of the paper to be produced, but because of the presence of lignin in the fibers, the quality can never be raised to the level of the chemical pulps. Groundwood pulp may be refined to improve the strength slightly, but the amount of increase possible is not great and the cost of treatment makes the operation unpopular. The groundwood pulping process, therefore, usually ends with screening the pulp.

PULPING CHEMISTRY AND PROPERTIES

The groundwood process just discussed removes the fibers from the tree by mechanical methods completely. The opposite end of the spectrum is represented by the *full chemical processes* in which the fibers are removed completely by chemical means. The full chemical processes, *kraft* and *sulfite*, remove the fibers from the wood by dissolving the lignin that holds them together in the tree. The mechanical pulp will be weaker and less permanent and will require more energy to produce, while the chemical pulp will be the opposite. There are many pulping processes that use a combination of chemical and mechanical energy and produce pulp with intermediate properties. The full chemical pulps will be discussed first, and then these combination processes can be better understood. The discussion of the chemical pulping processes will be divided into the chemistry of the processes first and then the machinery and processes used.

Kraft Pulping

The *kraft pulping* process has become the most common process for the production of full chemical pulp. The reasons for its success are (1) the strength of the pulp, (2) the versatility of the process in its ability to handle a wide range of raw materials and (3) the ready availability of a chemical recovery system. Kraft pulping enjoys the reputation of producing the strongest pulp available. This reputation has been earned by the fact that the damage to the fiber structure through chemical attack in the digester is reduced due to the nature of the chemicals used. The fact that many types of wood have been and are being cooked or pulped by the kraft process is due to the versatility of the process, and also to its tolerance to materials present in the wood that have caused problems with other processes. The kraft process will handle woods with high resin content, and in some cases the process can even be modified to allow the removal and conversion of these and other organic materials into useful by-products (tall oils, turpentine).

The kraft process is basically an alkaline cook, with sodium hydroxide (NaOH) being the primary cooking chemical. The first alkaline process was called the soda process and produced pulp by cooking chips in a sodium hydroxide solution with a pH of about 12. The wood, when it is cooked, will release acids into the cooking solution to reduce the pH during the cooking process. The NaOH can enter into several different reactions with the wood, some of which will also contribute to the consumption of the NaOH and the subsequent lowering of pH. If the pH is allowed to decrease during the cook, the result will be degradation of the cellulose and loss of pulp quality. The loss of quality may show up either in strength loss or color development. This quality reduction can be counteracted by increasing the initial caustic concentration in the cooking liquor. However, too high a concentration of NaOH at the beginning of the cook can also prove to be harmful to the chips. The solution to the problem was found through using sodium sulfide (Na_2S), which functions as a buffer or caustic donor. Equation (1) shows the equilibrium reaction, which demonstrates how the sulfide combines with water to contribute to the caustic concentration.

$$Na_2S + H_2O \rightleftharpoons NaOH + NaHS \quad (1)$$

The controlling factor in this equation is the relative concentration of the compounds on either side of the equation. If the caustic concentration is reduced, the reaction will proceed to the right to replenish the caustic concentration. The inclusion of the Na_2S in the cooking liquor helps control the pH throughout the cook and thereby prevents, or at least reduces, the degradation, or quality loss, of the resultant pulp. The equilibrium reaction in equation (1) shows that the caustic concentration remains suppressed as long as the pH is high, but the potential for regenerating or maintaining the caustic concentration as the pH is lowered is available until the Na_2S is also expended.

The sodium sulfide is beneficial as a buffer or caustic donor, but can also serve other functions in the cook. The sulfur is capable of reacting with the lignin to aid in its removal. The chemical nature of the lignin is such that it is a very large molecule that is made up of a large number of phenolic submolecules, or building blocks. In other words, there are several basic smaller molecules that have reactive points, usually more than one per molecule, that are capable of bonding with each other. The fact that there are many basic building blocks in the lignin and many points and ways for them to be attached to each other means that the lignin molecule is hard to characterize and cannot be represented by one chemical formula. The lignin in the tree is found in many different forms. The combining of the building blocks takes place along with the formation of the secondary cell walls during the fiber growth cycle. This combining of the building blocks not only causes the random or varying structure of the lignin, but also causes lignin to be situated throughout the fiber wall structure. The best term to describe the way the lignin permeates the fiber wall is interpenetration. It is an interpenetrating chemical structure. The caustic reaction with the lignin is primarily to break the bonds, or linkages, between the basic building blocks. The reaction of the sulfur with the lignin is for the sulfur to attach to the lignin fragments that have been broken loose. The sulfur attaches itself in such a way as to create acid groups or charged points that aid in holding the fragments in solution and thereby contribute to their removal from the wood or fiber structure. The substitution of the sulfur on the lignin compounds creates a molecule that therefore is more soluble in water than before. The molecules cannot be dissolved, however, without the presence of the sodium ion; it furnishes the counterion to stabilize the solution or dispersion of the lignin acids.

The chemistry of the kraft process can be summarized in the following manner. The sodium hydroxide contributes to the reduction in size of the lignin molecules or breaks the lignin down into its basic building-block components. The sodium sulfide contributes to the maintenance of the desired pH level and helps to buffer the reaction of the caustic with the wood to prevent or reduce damage to the pulp. The sodium sulfide also provides sulfur to react with the lignin building blocks, making them more soluble. Both the caustic and the sodium sulfide

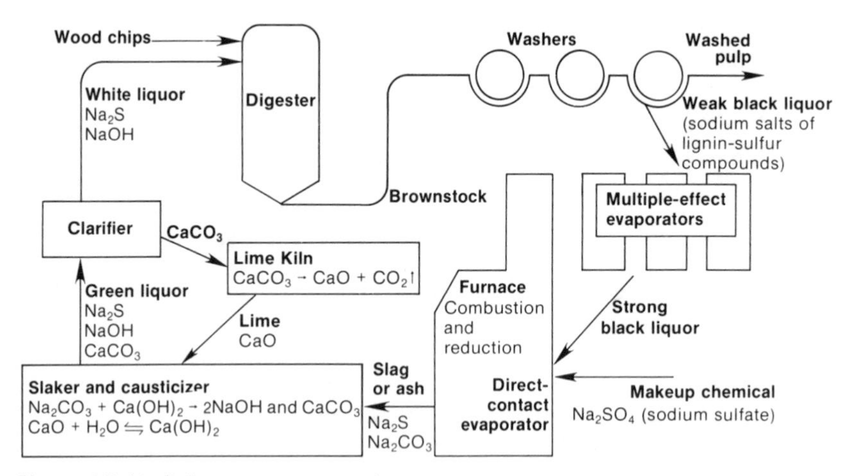

Figure 4.7. Kraft liquor recovery cycle

contribute sodium ions, which help in removal of the lignin reaction products from the wood.

After the pulping operation, the spent cooking liquor is removed from the pulp and burned to recover the cooking chemicals. A simplified diagram of the kraft recovery process is given in Figure 4.7. The kraft liquor recovery system consists of first thickening the spent liquor (black liquor) through several evaporation techniques, and subsequent burning of the thickened liquor in the recovery boiler. Not all of the sodium sulfide is recovered and some additional chemical must be added to replace the small amount that leaves with the pulp. The cheapest form of sulfur that can be added to the liquor prior to burning is sodium sulfate (Na_2SO_4). The burning takes place under both oxidizing and reducing atmospheres. In the reducing zone the sulfate is converted to sodium sulfide. The use of sodium sulfate as the makeup chemical in the recovery system has led to the other name for the kraft process—namely, sulfate pulp. Either name is proper and both refer to the same pulping operation.

One of the disadvantages of the kraft process is that small amounts of highly unpleasant-smelling sulfur compounds are emitted. Kraft pulpmills are constantly working to control these emissions; but the kraft process still produces a recognizable odor, even in compliance with air pollution regulations.

Sulfite Pulping

The *sulfite pulping* process is completely opposite to kraft pulping in some ways and similar in others. They are similar in the use of high temperatures and pressures during cooking and in the use of sulfur compounds to help remove the lignin. The sulfite process uses sulfur dioxide (SO_2) dissolved in water to produce an acid condition to help break down the lignin. The cooking liquor is prepared by burning sulfur in a controlled atmosphere to produce sulfur dioxide, which when dissolved in water forms a weak acid, which will react with the lignin. The reaction not only breaks the lignin into smaller parts, but also forms acids by the lignin and the sulfur combining to form molecules called lignosulfonic acids. These acids can be dissolved from the wood if there is a positive counterion or base present; therefore, the other part of the sulfite cooking liquor needs to be a base or positive ion. Calcium was originally the preferred base because of its low cost and availability. However, the presence of the calcium ion in the waste liquor causes problems. After cooking, the waste pulping liquor is thickened by evaporation until it is thick enough to burn. The burning can be controlled to give back a form of the original chemicals, which can be used to make new pulping liquor. However, if calcium is used as the base, the calcium can cause scale in the tanks and pipes used to thicken the liquor. The calcium forms calcium carbonate scale with the other products of the pulping reaction and can plug pipes quickly, requiring that the recovery operation be shut down and cleaned. Newer sulfite operations are being built to use sodium, magnesium or ammonia as the base, with fairly good results. The other problem with sulfite pulping is the presence of the sulfur dioxide which must be handled carefully because it is a toxic gas.

The properties of the sulfite pulp are also quite different from those of the kraft pulp. The fibers produced are considerably whiter and can be said to be cream colored, allowing them to be used directly in paper or board applications where high brightness is not needed. The fibers can also make paper that is softer or smoother than that from kraft pulps, which has led to the use of sulfite pulps in tissue and high-quality bond or writing papers. The other factor that has favored the use of sulfite in bond or writing papers is that the pulping operations leave behind fibers that have more pure cellulose in them. If the pulping operation is followed by bleaching, the resultant pulp is brighter and purer than kraft pulp and will give paper greater permanence than kraft. Because of the potential pollution and recovery problems, however, sulfite pulp is less favored than kraft and is currently on the decline as a major pulping operation. This decline is reflected clearly in Table 4.7.

FULL CHEMICAL PULPING OPERATIONS

Methods for cooking the chips can be divided into two basic types of operation: *batch* and *continuous*. As the names imply, the batch operations are carried out as sequential cooking steps and the continuous are carried out in a special tank that allows the chips to be fed in at one end and cooked pulp to be discharged at the other.

Batch Pulping

For the batch operation, the chips are loaded into a tank called a *digester*; the digester is sealed; the cooking liquor charged into the digester; the pulping operation carried out; and at the conclusion of the cook, the digester is emptied and refilled for the next cycle. To ensure a continuous flow of pulp for the subsequent operations, it is generally necessary for a mill to have several digesters, usually mount-

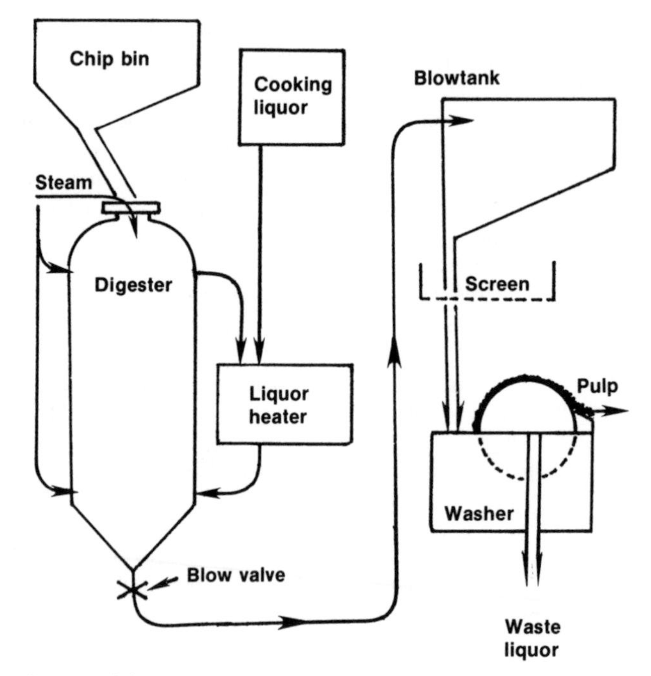

Figure 4.8. Batch pulping

ed side by side on the digester or cooking room floor and operated with one chip supply and one liquor supply system.

Figure 4.8 is a simplified drawing of a batch pulping operation. The chips would be delivered to the chip bin from the chip storage area following the necessary screening operations. When it is desired to charge, or fill, a digester, the lid is removed from the digester, the chute placed in position to fill it and the chips dumped into the digester. It is desirable to use a chip distributor to spread the chips out to ensure level filling of the digester and prevent bridging, or formation of dome-shaped piles, in the digester. If the chips do form a bridge or cone, when the liquor is pumped in, it may run off the top of the cone, much the same as rain runs off the roof of a building, preventing penetration of the cooking liquor into the entire pile of chips.

If it is necessary for the chips to be presteamed to heat them or to increase their moisture content, the presteaming would usually take place in the digester. Presteaming can very easily be accomplished in the digester by opening the steam valves leading to it and blowing live steam in among the chips. Presteaming can also be accomplished during the loading cycle by having the steam blown into the chip stream while the chips are being loaded into the digester. Blowing steam into the chip stream as it is being loaded into the digester will also help distribute the chips, in addition to increasing the moisture content and heating the chips. Heating the chips and the inside of the digester is another important function of the presteaming operation. If, for example, a load of cold chips is put into a cold digester and steam or hot liquor is pumped in, the steam will condense on the cold chips and the condensate will dilute the cooking liquor. Therefore, it is desirable to have some means of removing condensate created during the presteaming phase from the digester before the cooking liquor is added.

When the digester is filled, the chute is swung out of the way and the lid placed on top of the digester. The lid is usually a very heavy steel flange, which is secured to the top of the digester with large bolts and nuts. The cooking liquor can then be pumped into the digester. Usually, however, there will not be enough heat present in the liquor to make the entire digester and load of chips hot enough to carry out the cooking operation. Therefore it is necessary to heat the digester and its load during the cooking process. Steam can be used to heat the digester; however, the steam will dilute the liquor. Therefore, it is desirable in many cases to resort to external liquor heating. External liquor heating is accomplished by an arrangement of screen plates in the digester wall to allow the removal of liquor without loss of any of the chips. The liquor is then pumped to a heat exchanger, which will heat the liquor to the desired temperature and return it to the digester.

During the cooking operation, the temperature in the digester will rise to the desired degree and then the heater will be cut off or stopped. Increased temperature in the digester will cause the chips to give off steam and other gases, which will contribute to an increase in pressure inside the digester. The pressure that develops in the digester may become greater than that which would normally be associated with the temperature due only to steam pressure. This is especially true in the sulfite process, where sulfur dioxide is present in solution in the cooking liquor. Since sulfur dioxide is less soluble in hot cooking liquor than in cold, when the temperature of the cooking liquor is increased, sulfur dioxide gas is released and must be removed from the digester. Sulfite pulping operations therefore will have additional tanks, which are used to accumulate this sulfur dioxide gas as it is released from the digester. The sulfur dioxide gas thus accumulated is used to strengthen, or fortify, cooking liquor before it is put into the digester.

The temperature and times used in the digester vary greatly with the type of process and type of wood being cooked as well as with the strength, or amount of cooking, that is desired. Sulfite cooks, for example, may run to temperatures of from 125° to 160°C (250° to 320°F) with a maximum pressure of about 90 to 110 psi. The time for a sulfite cook may vary all the way from 6 to 12 hr. Kraft cooking temperatures will range up to around 170°C (350°F), but the time for the cook will usually be shorter than that for the sulfite pulp and may not last more than a few hours. However, if a very strong or complete kraft cook is desired, the time may be as long as 8 hr. It must be remembered that the main purpose of the cooking operation is to cause the liquor to penetrate the chips, dissolve the lignin and break down the chip structure. The liquor penetrates the chips partially by capillary action, but also due to the pressure that exists in the digester. As the cooking chemicals penetrate the chip, they react with the lignin in the chip or in the fiber walls and also with the cellulose in the fibers. If the cooking cycle is not allowed to last long enough, the chips will not be completely cooked. An undercooked batch will not break up easily, and rather than individual fibers, some chips or chip fragments will remain at the end of the cooking cycle. If, on the other hand, the cooking cycle lasts too long, the reaction between the cellulose and the cooking liquor will go too far and considerable degradation of the fibers will result. It is also possible that an overcooked pulp will become too darkly colored, making it difficult to bleach easily, or at all.

The reasons for concern about the size of the chips in earlier discussions now become more apparent. If we feed the digester a mixture of large and small chips, some chips in the mixture will be so large that they will never be completely cooked and small chips may be overcooked. The time, temperature and cooking liquor concentration therefore must be designed to suit the average chip size in the digester.

When the time has come for the cook to be finished, the top lid on the digester is kept in place, maintaining the pressure inside the tank. The blow valve at the bottom of the digester can be opened, and the pressure inside the digester is then used to push or blow the cooked chips from the digester through the pipe and into the blowtank or blowchest. The combination of release of pressure from the digester and impingement on the wall of the blowtank breaks down the chips into individual fibers. As the chips flow from the digester, it may be necessary to introduce steam to complete the blowing of the chips from the digester or it may be necessary to add waste cooking liquor to flush the remaining chips from the digester. When the digester is emptied, the lid is removed and the next cooking cycle may begin with the loading of new chips into the digester.

The blowtank originally was an open tank with a strainer or porous bottom to allow the spent cooking liquor to drain from the pulp. The use of an open blowtank presents a considerable pollution problem since the pulp, when it is released from the pressure of the digester to atmospheric pressure, will flash off steam and other volatile gases. These volatile gases will carry with them odors that may be undesirable, as well as chemicals that can be harmful both to humans and to plant life. The operation of a blowtank in a batch-type operation that subjects it to large blasts of pressurized chips intermittently has led to pollution problems and has been one of the contributing factors to the conversion from batch to continuous pulping by much of the industry.

The pulp goes from the blowtank through a screen, sometimes called a "bull screen," to remove knots and uncooked chips, and on to the pulp washer. The screen used in this position can be a drilled plate or a large-mesh wire screen, usually vibrated to facilitate passage of pulp and removal of oversize material from the surface. The knots, chips and uncooked pieces of wood removed from the stock on the bull screen can either be sent through the digester again or used as fuel in the waste liquor furnace.

Pulp Washing

Following the cooking and screening operations, it is necessary to remove the waste liquor from the stock to reclaim the liquor and also to produce high-quality pulp. A typical *stockwashing operation* is outlined in Figure 4.9. The rotary drum washer is designed such that the stock to be washed is introduced into a tank under the washing drum. The water in the stock passes through a wire screen on the surface of the washing drum, causing a pad of fibers to build up on the drum

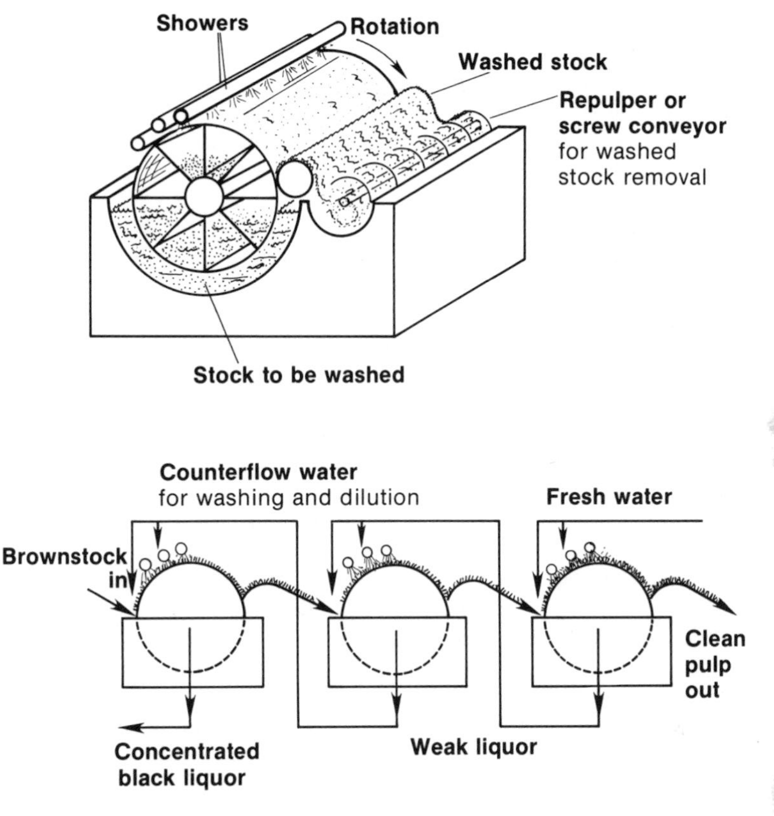

Figure 4.9. Basic washer design

surface. The pad of fibers is raised up out of the tank by the rotation of the drum and washed further by showers located above the drum. The washed stock can be removed from the drum surface, mixed with water and pumped on to the next operation. The vacuum drum shown in Figure 4.9 is divided into sections to allow the use of a partial vacuum inside the drum. The vacuum must be increased as the thickness of the pad increases, to offset the reduced flow rate due to the thickness of the pad. As the drum continues to rotate, the vacuum may be continued to remove water from the bottom of the pad as showerwater is being sprayed on the top. Such operations are called *displacement washing.* The soiled water in the pulp is displaced by the cleaner water from the showers.

The counterflow washing principle is demonstrated in the lower part of Figure 4.9. The dirtiest stock (brownstock in the illustration) is introduced into the first washer on the left and is washed with the dirtiest water, which was obtained from the washing operation in the second washer. The countercurrent flow of the washwater and the stock allows us to minimize the amount of fresh water required, and also increases the concentration of the chemicals in the wastewater removed from the first washer.

From this point in the treatment of the pulp, there is no difference between batch and continuous cooking operations. The treatment of the pulp after this point depends on what kind of papermaking process it will go into, and therefore the description of subsequent treatments belongs more properly in the discussion of stock preparation and papermaking operations in Chapters 5 and 6.

Continuous Pulping

The *continuous digester* will accomplish the same cooking and breakdown of the chips into individual fibers as was accomplished in the batch-type digester. The obvious difference is in the continuous nature of the operation. The continuous digester must be fitted with some form of mechanism to allow the continuous introduction of chips and removal of cooked chips from the bottom of the digester. We will see that the operation becomes more complicated than this simple straightforward representation in that the continuous digester has been modified to allow washing of the chips while still in the digester.

Figure 4.10 is a diagram of the operations and machinery involved in the continuous pulping scheme. The chips again are brought to the digester area chip storage through the screening operation. Screening and maintenance of chip moisture must be performed for continuous digester operations the same as for batch digester operations. The need to control the cooking time and temperature is the same for continuous digester operations as for batch operations. The time in the digester is controlled by the flow rate of chips to the digester. In other words, if the digester has a capacity of 100 tons of chips and we pump 100 tons of chips through per hour, the chips will be in the digester for an average of 1 hr. At the point where the chips leave the chip bin, they are put under a low to medium pressure by using steam to blow the chips from a low-pressure meter into the presteaming vessel. The low-pressure meter is a feeder, similar to a revolving door, that has compartments to allow the introduction of chips on one side. As the meter rotates, the compartments filled with chips are carried to the other side, where the chips can then be blown or dumped into the steaming vessel. The presteaming vessel performs just that function—the chips are presteamed in this tank. The size of the tank again is determined by the flow rate of chips and how long it is desired to have the chips in contact with the steam. At the exit end of the presteaming vessel, there is a high-pressure valve. This valve is similar to the low-pressure meter except that much closer tolerances must be used in its design to maintain the pressure in the high-pressure line. The chips are carried into the pockets in the high-pressure valve by the steam and condensate that are present in the presteaming vessel. As the valve rotates, the pocket (at this time filled with chips) passes across an opening where liquor is pumped in. The liquor pumped into the valve pushes the chips out of the pocket and carries them on toward the digester. The design of this high-pressure valve then is more similar to the valve or stopcock in a chemical burette, which has a hole through the center plug that will line up with pipes on either side of the center plug as it rotates. As this hole through the center plug lines up with the chip line coming from the presteaming vessel, the hole fills with chips. As the hole lines up with the pipes in the high-pressure liquor circulation line, the chips are forced out and carried up to the

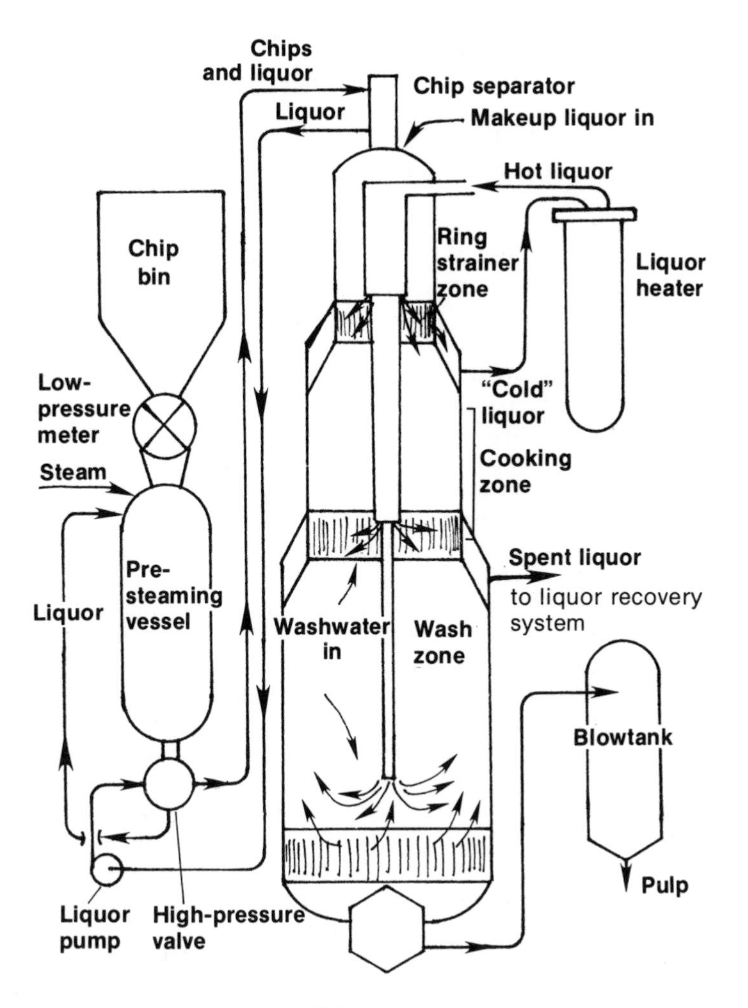

Figure 4.10. Continuous cooking

chip separator at the top of the digester. The chip separator may be similar in design to the high-pressure valve or it may be a purely physical screening device that will separate the chips from the liquor, allowing the liquor to be carried back to the high-pressure liquor pump. The liquor being used at this stage is full-strength cooking liquor.

Once the chips are dumped inside the digester, they build up on top of other chips already present in the digester, forming a large pile of chips extending all the way from the bottom to the top of the digester. If the rate of introduction is equal to the rate of removal, then we have a static amount of chips in the digester and the pile will remain the same size. It is therefore possible to have a pile, or plug, of chips that is slowly settling toward the bottom of the digester, and at the same time to be pumping the pulping liquor through the chips as the chips settle down through the digester. As the chips settle toward the bottom, liquors are pumped in through a collection of pipes coming down the center of the digester and removed through the ring strainers at the outside edge of the digester, as shown in Figure 4.10.

More specifically, makeup liquor will be added at the top of the digester to compensate for the liquor that is removed by the chip separator and returned to the liquor pump. As the chips settle down through the digester, the liquor begins to penetrate them. The rate of reaction between the cooking liquor and the chips is dependent upon the temperature. Therefore the cooking zone is indicated as being in a central section of the digester, after the temperature has been raised to the desired level. The cooking zone follows the first ring strainer zone, where the liquor is pumped out through the heat exchanger to increase its temperature and then pumped back into the digester. This kind of liquid replacement is used very extensively in continuous digesters. The chips, you must remember, are not removed with the liquor, but remain in the digester, settling slowly toward the bottom at all times. The hot liquor is pumped in at the first opening in the nested tubes coming down the center of the digester, as shown in Figure 4.10. This hot liquor being pumped in forces the cooler liquor toward the outside edge, where it is removed by the ring strainer located all the way around the outside edge of the digester. The hot chips and liquor then settle slowly through the cooking zone without any other liquor being pumped in or out. At the end of the cooking zone, washwater or spent liquor is pumped into the digester through the nested pipes, forcing the cooking liquor out through the outside strainer and on to the liquor recovery system.

The chips at this time are fully cooked and can be washed in the same digester by the introduction of washwater through the inside pipe in the nested pipes or through strainer plates in the bottom of the digester. Enough water is introduced at the bottom of the digester to force the water up through the chips and out at the spent liquor strainers mentioned previously, as shown in Figure 4.10. It is possible to use hot water in these washers to maintain the temperature and pressure that was used in the cooking zone. It is also possible to use cooler water to reduce the temperature of the chips as they settle toward the bottom of the digester. It is therefore possible to operate a continuous digester such that chips enter the top of the digester at temperatures greater than 100°C (212°F); then, by removing liquors to the heaters and pumping the hot liquors back into the digester, the temperature may be raised to about 160°C (320°F), where it is held through the cooking zone. The wash cycle then can reduce the temperature of the chips to around 100°C (212°F) or below. The removal of chips from this kind of digester is far easier than from the batch-type digester. We no longer have the very hot chips, which will flash off steam and other gases, nor do we have the need to pump or release the cooked material into an open tank. Because we are producing a continuous flow of chips or perhaps washed pulp, it is possible to pump them into a closed tank without fear of the tank collapsing when the pressure is reduced at the end of each cook, as in a batch operation. The blowtank design will be very similar to that of a cyclone separator where the chips or pulp

Table 4.4. General Description of Mechanical Pulp

Name of process	Strength of pulp	Pulp usage	Final products	U.S. Capacity 1988 (000 short tons)	Number of pulp mills in U.S. 1989
Groundwood (GWP)	Low	Usually mixed with chemical pulp.	Newsprint, magazine, book, catalog, tissue/towel, boxboard, building, insulating board	3,104	71
Refiner mechanical (RMP)	Low but higher than GWP	Blended with GWP or chemical.	Newsprint	335	7
Thermomechanical (TMP)	Low but higher than RMP	Same as above.	Newsprint, publication papers, fiberboard	3,207	23
Chemithermomechanical (CTMP)	Higher than TMP but lower than kraft	Same as above above.	Newsprint, publication papers, tissue/towel, bleached board, fluff	30 (2,500 in Canada)	1 (15+ in Canada)
Pressurized groundwood (PGW)	Nearly as high as TMP; higher than groundwood groundwood	Same as GWP	Supercalendered papers, coated groundwood	510	4

Source: *Pulp & Paper North American Factbook*, p. 271.

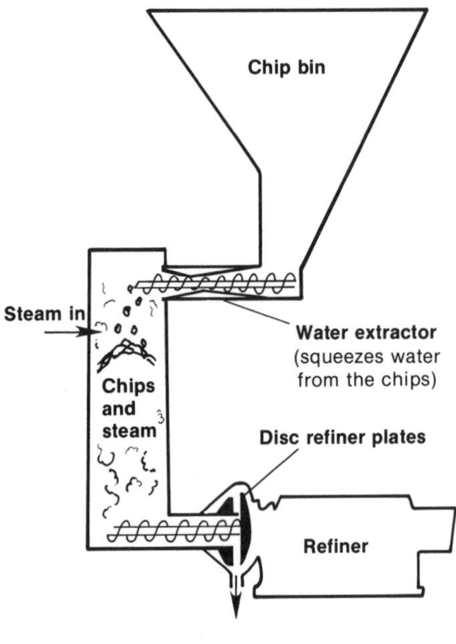

Figure 4.11. High-yield mechanical pulp process

mechanical pulps are given in Table 4.4. These processes all use primarily softwoods (spruce, balsam, pine and hemlock) but may also use hardwoods (aspen or cottonwood).

More detail on the operation of the disc refiner is given in Figure 4.12. It can be seen that the chips are fed into the refiner by means of a screw feeder on the left side of the drawing. The chips are fed between two discs, one of them stationary and the other rotating at a high rate of speed. The surfaces of the plates are covered with raised bars of varying size, as indicated in Figure 4.12. The bars are closer together at the outside edge of the plate than at the middle, but the actual angle is exaggerated in the drawing. As they pass between the plates, the chips are broken and ground down eventually to fibers, requiring that the clearance between the plates be reduced. Some processes are able to completely defiber the chips in one pass through one refiner, while others require the use of two refiners in series to complete the job.

The next step in the movement toward chemical pulps would be the use of a chemical treatment in the processes already described, leading to the names *chemithermomechanical pulp* (CTMP), *thermochemical mechanical pulp* (TCMP) or *chemirefiner mechanical pulp* (CRMP). The chemical treatment may be a soak period for the chips before or after the water extractor shown in Figure 4.11, or may be introduced after the refiner and before the addition of a second refiner. Many different processes are possible, leading to the variety of names used for this group of processes.

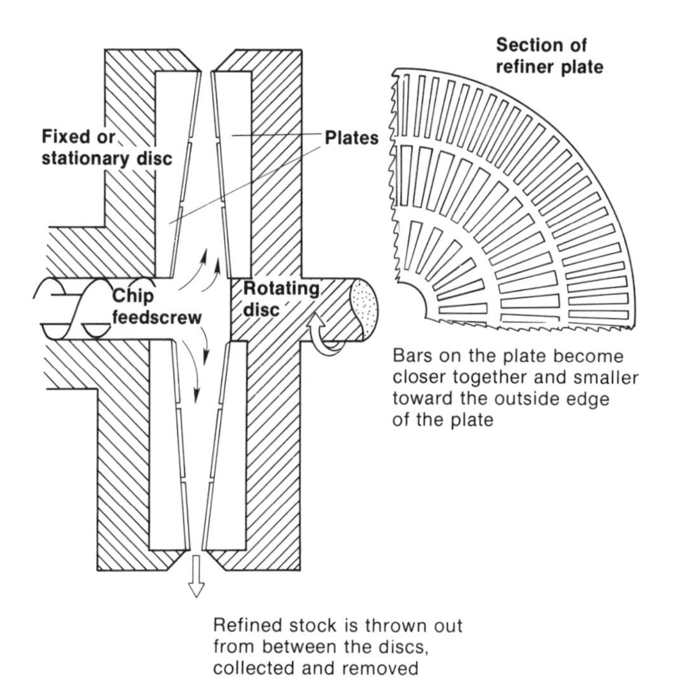

Figure 4.12. Disc refiner detail

The use of a pressurized soak and perhaps even a continuous digester ahead of the refiner brings us to a family of processes known as semichemical pulps or just high-yield pulps. The most common of these is the *neutral sulfite semichemical* (NSSC) *pulp.* NSSC originally used a sulfite liquor neutralized to a pH of about 7 with waste liquor, a short low-temperature cook and mechanical treatment. The process still can be found operated that way, but it has also been modified in a number of applications to use all kinds of chemical liquors, temperatures and pressures.

Most of the high-yield pulps up through tandem TMP in Table 4.3 are being used as replacements for groundwood and are being used with little or no bleaching. There is some interest in producing bleached CTMP, which would be called BCTMP, for use as a replacement for bleached chemical pulps and also to be sold as a market pulp. Although there is some being used, economics have not favored this trend. Some of the TMPs are being used as a filler in boxboard, but only in one or two mills. The properties of these pulps are close to those of groundwood but are necessarily variable and dependent on the actual process used. The NSSC has been the pulp of choice for corrugating medium, using primarily hardwood chips, very little washing and no bleaching and relying on the presence of some of the spent liquor and lignin to give the paper its characteristic stiffness after it has been fluted and is assembled into the corrugated board.

PULPING OF SECONDARY FIBERS

The pulping of secondary fibers, or wastepaper, is considerably more simple than the methods discussed for the pulping of wood. The main job to be accomplished in pulping wastepaper is simply the separation of the web or sheet into its individual fibers. The project is complicated by the various chemical treatments and coatings the paper may have received in processing, and if the paper has been printed, the printing ink could cause difficulty. The removal of the ink is considered as a separate treatment called deinking. In either case the fibers must be separated first before any further treatment can be carried out.

Pulping Methods

The most common device used to defiber secondary fibers is the *pulper*, shown in Figure 4.13. This device is loaded with water and then the dry wastepaper is fed in, usually from a conveyor belt. Enough wastepaper is added to bring the solids content (called *consistency*) up to at least 5% to 6%. The higher the consistency, the better, because the machine relies on a certain amount of rubbing between the fibers or pieces of paper to do the job. The major part of the job of breaking the paper into fibers is accomplished by the rotor in the bottom of the pulper, which spins at a high rate of speed and tears the paper into small pieces. The rotor also must be designed to cause the pulp to move around in the tank, to allow all of the charge to flow past the rotor so that it may be broken down. Once the bigger pieces have been broken down and the consistency increases, the rubbing together of the fibers or fiber hunks helps break them down into individual fibers.

Figure 4.13. Pulper for secondary fibers

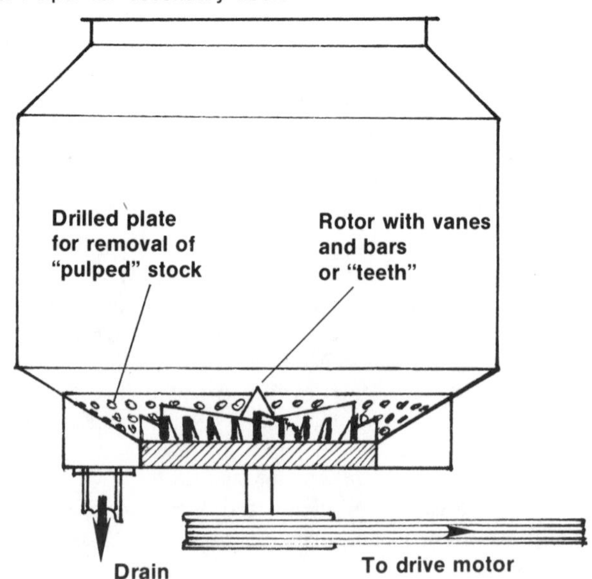

Stator/rotor for use in a pulper with the rotor installed in the side wall. Cutaway portion is intended to show the extraction zone behind the rotor.

The pulper can be operated in either a batch or a continuous mode. For continuous operation, the pulper must be fitted with a drilled plate or screen in the bottom. The holes in the plate may be as big as ½ in. in diameter, but they will still be small enough to reject pieces of paper and accept only fibers. For continuous operation, a steady flow of stock will be drawn off the bottom of the tank through the strainer plate, while water and dry wastepaper are added either constantly or intermittently to maintain both the level of stock in the tank and the consistency of the stock. Stock removed from the pulper will normally be passed through some form of screening, which will remove pieces of paper not broken down. Sometimes disc refiners similar to those used in high-yield pulping operations can be employed to help break up small pieces of paper not broken down by the pulper.

Some wastepaper will have been given either chemical treatment or coatings that will make it too difficult to break down without more help. Simple heating of the water to about 65°C (150°F) is common to help break down the paper. Chemicals such as sodium hydroxide and/or soaps and dispersing aids can also be added. However, although soaps and dispersants can help break down the paper, they also may cause foam in the pulper or in later handling steps. And sodium hydroxide can cause discoloration of the fibers, necessitating bleaching to restore the color of the fibers.

Deinking Operations

If printed paper is reclaimed for reuse in the manufacture of white paper, the ink must be removed by some form of cleaning operation. If the paper is coated and the ink is adhering only to the coating, the ink can simply be washed off the paper. In most cases, however, the printing has been done directly on the fibers and the process is not so simple. The *deinking operation* actually begins in the pulper with the selection of the chemicals to be added there. The chemicals will not only help to break up the paper but may also help to disperse the ink and remove it from the fibers. Most deinking operations use different types of washing equipment along with special chemicals to disperse the ink and make it easier to remove. The washers may be simple drum washers like those used in the chemical pulping and bleaching operations. More frequently, we find the dilute stock being passed over inclined screens or slotted plates, which allow water and the dispersed ink to pass through while rejecting the fibers and allowing them to be concentrated on the surface. More recently, foam flotation techniques have been adapted to the job of deinking. Foam or froth flotation processes operate on low-consistency stock suspensions, with the ink being collected by the foam and removed from the fibers. Either process can operate effectively if the wastepaper used is suitable for deinking.

Not all paper can be deinked easily or economically. The largest obstacle to increased use of wastepaper is the cost of collecting and sorting it. Many forms of paper can be reused, but each requires a slightly different treatment to be used most effectively. Contamination of one usable type of wastepaper by another type of paper, which also might be perfectly usable if treated by another process, can cause considerable difficulty. The inclusion of some plastics or pressure-sensitive adhesives in wastepaper can make it practically unusable. Although wastepaper may be readily available in mixed form, the cost of sorting, coupled with the cost of pulping and deinking, raises the overall cost to nearly the same as virgin fiber. Unfortunately, the strength of wastepaper or secondary pulp is not as good as that of virgin fiber and therefore cannot compete at an equal cost. Certain grades of paper that can be collected in printing plants and that all have the same type of printing, or very little printing, can be sold for a better price and are usually worth the additional cost to the paper mill that can use them.

The use of secondary fiber has been emphasized greatly in recent years, but is not new to the paper industry. Wastepaper has been a viable source of fiber since the very beginning. The difficulties in recycling stem from obtaining suitable wastepaper for the particular application; removal of the impurities in the wastepaper and disposing of this waste stream; and performing all of these operations at a cost that is competitive with using virgin fiber sources.

Comparative world consumption data presented in Table 4.5 shows some of the complexity of the world situation. Although we in North America are consuming a tonnage similar to Western Europe, our recovery rate lags behind these and many other countries. The key to the situation lies in the cost and availability of other fiber sources. In Japan, where wood is imported from other countries, the economic incentive to recycle is greater than in North America, where wood is still quite abundant. We therefore find that considerable wastepaper is exported from the United States to Japan and other countries. If we compare these consumption numbers with the per capita consumption data of Table 1.1, the potential for greater wastepaper utilization can be seen. As the demand for paper grows in the countries with low per capita consumption, the wastepaper sources for recycling are not present. We have seen problems arise from the depletion of forests in many regions of the world and considerable opportunity for the use of exported wastepaper is expected to develop.

If we consider the wastepaper available for recycle, we find that there are over

Table 4.5. World Wastepaper Consumption by Geographical Area in 1988

	Wastepaper consumption (1,000t)	Share of total world consumption (%)	Wastepaper recovery rate (%)
North America	19,778	26.5	29.4
Western Europe	19,811	26.6	35.8
EFTA countries	2,967	4.0	40.9
EEC countries	16,844	22.6	35.1
Eastern Europe	4,868	6.5	29.0
Oceania	702	1.0	24.8
Latin America	4,757	6.4	33.6
Japan	12,538	16.8	48.0
China	3,084	4.1	20.4
Rest of Asia	8,369	11.2	34.2
Africa	679	0.9	16.5
WORLD TOTAL	**74,586**	**100.0**	**32.7**

Source: A. C. Veverka, 1990.

70 grades of stock listed in "Circular PS-8a5 Paper Stock Standards and Practices" issued by the Paper Stock Institute of America (330 Madison Ave, New York, NY 10017). Of these over 40 grades are being shipped on a regular basis, but only a handful accounting for the major tonnage consumed (Table 4.2). The largest tonnage is old corrugated containers (OCC), which can be further classified to scraps collected at the corrugating plant through the used containers collected from markets. The latter is likely to be contaminated with other packaging materials and therefore is less desirable and more costly to process. News is the second largest category. It can also be broken into cleaner grades from printing operation waste to old newspapers collected by local service and charitable organizations. Mixed is office waste, contaminated industrial waste and so on. It is highly contaminated and not too valuable as a raw material, since it is quite costly to clean up but does not necessarily contain high enough quality fibers to justify the cost. Mixed is the lowest grade, and is as likely to go to landfill as recycling. Other specific grades have been used where a stream of either clean or consistently contaminated waste with high-quality fiber can be guaranteed to justify the development and construction of a mill to process it. Several regions have established successful newsprint recycling operations by setting up collection systems to guarantee a source of old newspapers at a cost suitable for the development of a mill to regenerate recycled news.

A summary of the current recycle situation is shown in Table 4.6. Containerboard (corrugated) is not only a major user of wastepaper but becomes the largest source of fiber for recycling. Part of the reason for this situation is the fairly high waste generated in the box plant; the other part is the high-quality fiber available in the waste generated. News is a rather specific product and source that has been suitable for collection and recycling in major metropolitan areas, as discussed earlier. Combination boxboard has been made from 100%

Table 4.6. Summary of Recycle Operations

Recycle containing grade	Wastepaper source
Containerboard—linerboard and medium	OCC[1], boxplant clippings, mixed
Groundwood—News	News, special DI[2] news
Folding boxboard (combination boxboard)	OCC, news, mixed (hard white clippings for top liner)
Tissue	News, DI, mixed, PS[3]
Printing and Writing	DI, (mostly printer's waste)

[1]OCC–Old corrugated containers
[2]DI–Deinking grades
[3]PS–Pulp substitutes

recycle fiber for many years and can use a variety of wastepaper streams in the filler or "chip" layers. However, the grade is not completely tolerant of contaminants and cannot use mixed grades or grades contaminated with plastic materials. Plastics tend to soften in the drier section and stick to the driers, carrier rolls or adjacent wraps of paperboard in the reel. These contaminants are referred to as "stickies" because of this behavior and are very undesirable. Tissue has become one of the major focus areas for the increased use of recycle fiber and is able to use quite a variety of fiber sources, but also is sensitive to the stickies problem. Printing and writing grades require clean waste with good fiber content and waste that can be deinked economically. Relatively clean waste collected as a by-product of printing operations (edge trim and similar scraps with little ink are the best) or waste with a consistent type and amount of ink, which allow a specific deinking process to be developed, are preferred.

BLEACHING

Both full chemical pulping operations and the high-yield processes leave the pulp too highly colored to be used in making white paper. Unbleached groundwood and sulfite have been and are being used in newsprint, but the brightness of these papers is not really too high. Furthermore, due to the presence of lignin, these papers do not have any degree of permanence and yellow easily. *Bleaching* not only improves the whiteness or brightness of the pulp, it improves the permanence of that whiteness. Bleaching can therefore be shown to perform two functions. The improvement in permanence is a result of the removal of lignin from the fibers. Therefore, bleaching can be seen as a continuation of the purification that begins in the pulping operations. The purification aspect explains the loss of weight experienced by the pulp during bleaching. Depending on the purity of the pulp before bleaching, the loss may be great or small. If it is known that the pulp is to be bleached to a high brightness, it is common to use a strong pulping cook to deliver a purer pulp to the bleaching operation. The removal of impurities also indicates the need for washing as an integral part of the total bleaching sequence. The pulp is normally subjected to washing immediately following bleaching to remove both the spent bleach liquor and the impurities.

Bleaching operations primarily depend on chlorine and chlorine compounds. Depending on the conditions of use and the needs of the pulp, chlorine is used in at least three different forms. Chlorine gas dissolved in water to a pH of about 2 is used as a common first stage of bleaching. The dissolved chlorine gas reacts with the lignin remaining in the pulp and creates a lignin acid, which can be dissolved from the pulp in subsequent stages. The pulp will be washed following this *chlorination* stage, and then sent to what is called an *extraction* stage. The extraction is accomplished by using a strong solution of sodium hydroxide, powerful enough to have a pH of about 12. The sodium hydroxide breaks down the lignin molecules and removes them from the fibers, using the sodium ion as the counterion to draw the negatively charged lignin acid from the fibers. It is hard for a casual observer to believe that chlorination and extraction are truly bleaching operations. The pulp will become yellow colored following the chlorination and brown colored following the extraction. Washing is necessary after extraction to remove the impurities, but the pulp will still require further bleaching to make it white. The remaining bleaching operations or stages are commonly oriented more toward removing color than impurities. Sodium hypochlorite, familiar to most as household bleach, is the most common stage used following chlorination and extraction. Treating the pulp with one or two stages of hypochlorite bleaching, and the wash following each bleach, will raise the brightness to fairly good levels. However, we do reach a point of diminishing returns where further bleaching with hypochlorite will not improve the brightness enough to justify the cost. Peroxide bleach is therefore a common bleach stage used at this point. The peroxide has the advantage of providing better permanence, in that pulp bleached with a final peroxide stage is less likely to yellow later. Another bleaching chemical that has become quite popular is chlorine dioxide. Chlorine dioxide can be used early in the bleaching sequence to help in purification or it can be used as a final bleach stage to give the pulp good permanence. It is common for highly bleached pulps to be subjected to several stages of bleaching. A common sequence for high-brightness pulp would be: chlorination, extraction, hypochlorite, perhaps another hypochlorite and either peroxide or chlorine dioxide. Washing stages would be inserted between each bleach and at the end.

The equipment used for the bleaching operations consists primarily of closed tanks into which the pulp is pumped in water suspension after being mixed with the bleaching chemicals. The pulp is carried by water throughout most of the bleaching operations. The chlorination stage is usually carried out at fairly low consistencies and temperatures. A consistency of 3% to 4% will be quite fluid and will flow freely; therefore chlorination is usually done at about 3% consistency. The washing is carried out in rotary drum washers similar to the ones described in connection with the washing that follows the pulping operations. The pulp coming off the drum washer will usually be about 6% consistency, which is high enough that it will not flow readily and looks as if it would need to be shoveled to make it move. Actually, there are pumps that can deliver 6% consistency pulp to thickeners, where the consistency will be raised even higher. The extraction stage can be performed at consistencies from 6% to 12% by raising the consistency and then diluting it again by the addition of the sodium hydroxide solution.

Subsequent bleaching stages are normally carried out at high consistencies. The higher consistency reduces the amount of dilution of the chemical, allowing less chemical to be used, and further saves energy by eliminating the need to heat excess water. The peroxide stage is specially suited to high consistencies because the peroxide can still function well when released as a gas in the bleach tower.

Not all pulps are bleached with multistage bleach sequences. Groundwood and secondary fiber pulps frequently receive only a one-stage bleach of either hypochlorite or peroxide. Secondary fibers are generally pure enough not to need the purification possible with a combination of chlorination and extraction. Groundwood and other high-yield pulps have too much lignin to be subjected to chlorination and extraction. Such treatment would lower the brightness sufficiently to require extensive additional bleaching, and the final yield would be so low that the advantage of using these pulps would be lost.

All of the bleaching operations discussed to this point have been what are classified as oxidizing bleaches. These are other bleaches that function in the opposite manner, as reducing agents. Sodium hydrosulfite is a reducing bleach that has been successfully used on groundwood and secondary fibers. These bleaching chemicals work well and can produce pulps of similar properties to the oxidizing bleaches.

WOODPULP PRODUCTION

Table 4.7 shows a breakdown of the U.S. production of the kinds of pulps described in this chapter. It should be noted that about 60% of the bleached pulp

Table 4.7. U.S. Woodpulp Production, 1970–1988

	1970	1975	1980	1985	1988
			(000 short tons)		
TOTAL: ALL GRADES	**43,546**	**43,084**	**46,024**	**54,147**	**61,210**
Sulfite, total	**2,344**	**1,951**	**1,673**	**1,560**	**1,560**
Bleached	1,931	1,594	1,324	1,364	1,365
Unbleached	413	357	349	196	195
Sulfate, total	**29,472**	**29,213**	**35,009**	**42,136**	**47,983**
Bleached and semibleached	13,472	13,844	17,236	22,626	26,302
Unbleached	16,217	15,369	17,773	19,510	21,681
Groundwood	**4,404**	**4,351**	**4,324**	**5,251**	**5,943**
Semichemical	**3,297**	**3,201**	**3,652**	**4,026**	**4,357**
Special alpha/dissolving	**1,705**	**1,583**	**1,366**	**1,174**	**1,367**
Defibrated/exploded	**2,105**	**2,785**	**n.a.**	**n.a.**	**n.a.**

Source: American Paper Institue and Bureau of the Census, as given in *Pulp & Paper North American Factbook*, 1980, 1984, and 1989.

Note: Some of the data for earlier years is not directly comparable due to changes in classifications.

made in North America is used on-site to make paper and paperboard. The remainder is machine dried for use at a producer's mill at another location or sold as *market pulp*, woodpulp produced for sale to another company in the same country or abroad.

Chapter 5 describes the process of pulping machine-dried pulp for manufacture into paper and paperboard. As can be seen from Table 4.7, most of the pulp produced in the USA is used in paper or paperboard manufacture. *Dissolving* or *special alpha* pulps are used in the production of man-made fibers (rayon and acetate), films (cellophane), plastics and chemicals. Detailed discussion of such products, which belong to the textile and chemical industries, would be outside the scope of this book. They are briefly mentioned here, however, because they represent yet another realm of products related to the pulp and paper industry.

REFERENCES

Britt, K. W., *Handbook of Pulp & Paper Technology*. New York, Van Nostrand Reinhold, 1964 (revised 1979).

Clark, J d'A., *Pulp Technology and Treatment for Paper*, Second Edition, San Francisco, Miller Freeman, 1985.

Koffinke, R. A, *Tappi Journal* 63 (11):51 (1980).

Kocurek, M. J., *Pulp & Paper Manufacture*, Third Ed., Vols. 1-5 The Joint Textbook Committee of the Paper Industry (TAPPI, Atlanta, and CPPA, Montreal), 1990.

Veverka, Arthur C., "The Cost Competitive Aspects of Recycled Fiber Usage," in *Proceedings of Wastepaper I: Demand in the 90's*, San Francisco, Miller Freeman, 1990.

5 Stock Preparation

The earliest form of mechanical treatment of fibers consisted of throwing old rags in a stone pit, allowing them to rot a bit and then pounding them with a stone. This device was called a *hammermill,* a name still in use as a tradename for a brand of typing or bond paper. The hammermill was replaced by the invention of the hollander beater in about 1700. For the next two centuries, the beater was used to perform all the operations discussed in this chapter. For this reason, the area devoted to these operations may still be called the *beater room* even though the mill may not have a beater.

OVERVIEW OF THE BEATER ROOM AND STOCK PREPARATION OPERATIONS

The *beater* is an oval-shaped tub with a dividing wall running the long dimension of the tub, but not all the way to the ends, as shown in Figure 5.1. On one

Figure 5.1. The beater

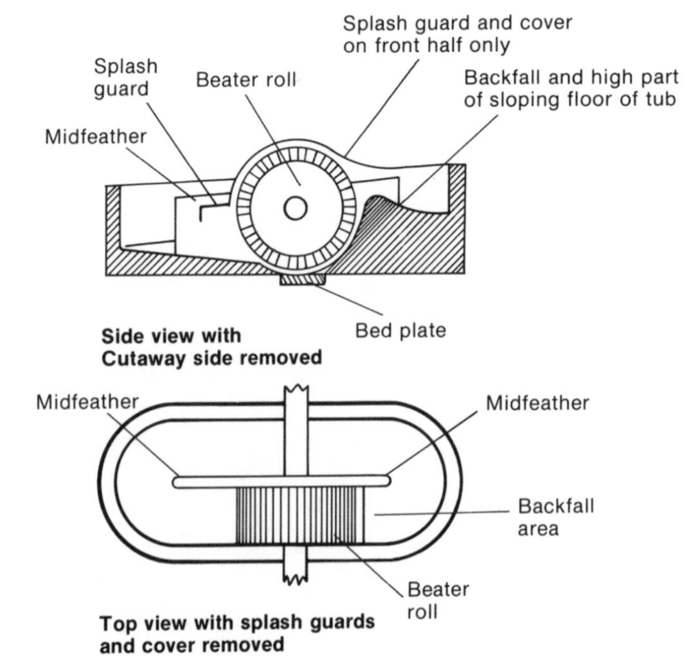

side of this wall, called the *midfeather*, is a large wheel with bars parallel to the wheel axis on its surface. Under the wheel or roll is a stationary bed plate with bars on its surface that are parallel to the bars on the roll. The roll is rotated, causing the stock in the beater to be rotated around the inside of the tub, and to pass under the roll. With a large clearance between the roll and the bed plate, there is little or no action on the individual fibers, and the beater can be used to break up paper or dry sheets of pulp, an operation sometimes called *pulping*. Once the stock has been broken down into original fibers, the roll can be lowered until it is close enough to beat on the fibers. The entire *furnish*, or mixture of all of the ingredients needed for making any type of paper, can be mixed together in the beater. With increased production rates and the demand for more streamlined, continuous operations, the beater was not able to keep up, but all the operations it once performed still must be carried out by its replacements.

All operations necessary for preparing the fibers, and included in the beater room or stock prep area, are summarized in Figure 5.2. Starting in the upper left corner, these operations begin with pulping and proceed through refining, metering and blending to dilution and cleaning. Refining is the most important of these operations and requires considerable treatment; it will be discussed later in this chapter. The other operations are rather straightforward and will simply be discussed briefly.

The *pulping* operation performed here resembles the chemical pulping operations discussed in Chapter 4 only in the general overall goal of the operation. The

Figure 5.2. Flow chart of stock preparation operations

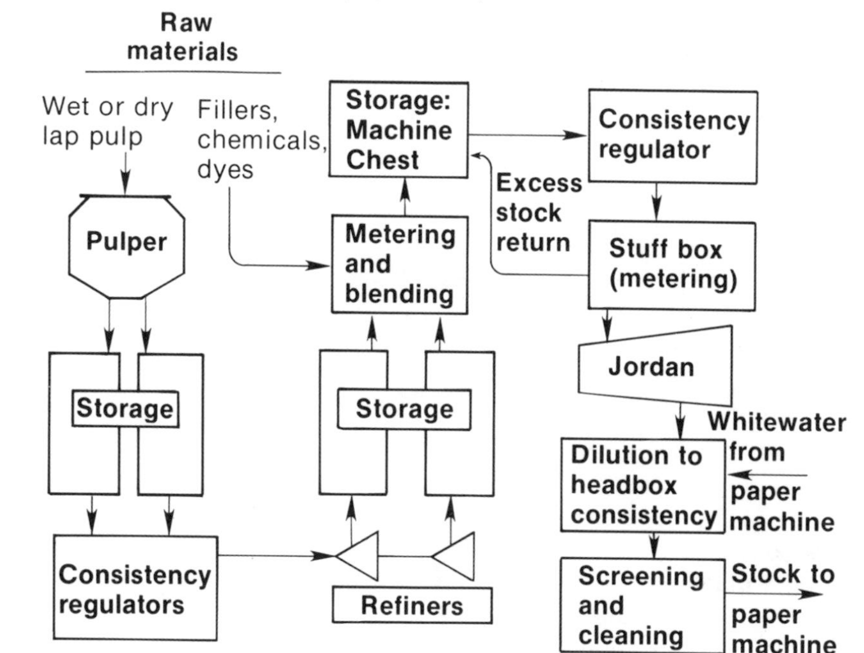

goal is simply to liberate the fibers from the raw material being used by the process. In the case of chemical pulping, that raw material was trees. In the case of a papermill not able to get its fibers directly from a pulpmill in a water suspension, the fibers must come from another form. The common form is called *dry lap pulp*. The fibers are liberated from the wood by a pulpmill; then they are formed into thick sheets of paper and bundled into bales. The pulp is not dried completely, but retains about 20% moisture to prevent the fibers from bonding to one another or collapsing. These bales of pulp can be sold by the pulpmill to papermills all over the world. Dry lap pulp is both a valuable product for some pulpmills and a valuable raw material for papermills.

One type of pulper is shown in Figure 5.3; it is representative of the pulper described in Chapter 3 to be used for the pulping of wastepaper. The pulper is equipped with rotating blades in the bottom, which serve to break the pulp into individual fibers. It may be operated as a continous pulper, with water and bales of pulp being added to maintain the desired level or volume and defibered stock being removed through the strainer plate in the bottom. The pulper is more likely, however, to be operated as a batch operation, being used to pulp up different types of pulp and send them to separate storage tanks. It is desirable to refine hardwood and softwood pulps separately, so separate storage and treatment facilities are needed until the blending stage, when the two are combined.

From storage, the pulp goes to the *refining* equipment by way of consistency regulators. We will see later that the consistency at the time of refining is impor-

Figure 5.3. Pulper

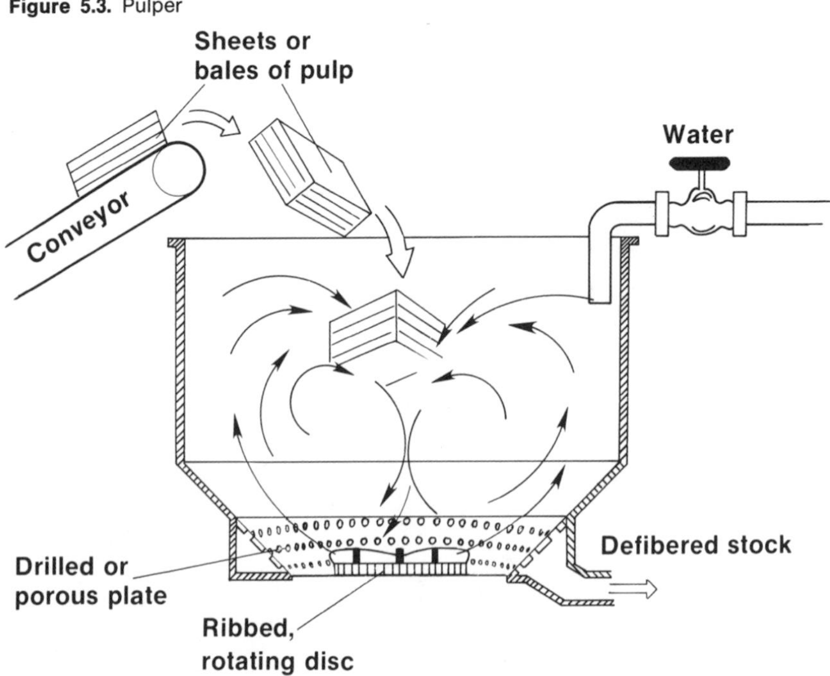

tant and must be controlled. Consistency regulators basically measure the viscosity, or resistance of the stock to flow, which is a function of the consistency. In order for the consistency regulator to correct the consistency to the desired level, the consistency must be higher than desired since we can only add water at this point. The consistency regulator is installed in the pipe line in most mills and will automatically measure the consistency, adding the water needed to adjust it to the desired level.

As with all other operations in papermaking, mills tend to use different types of refiners, and will use them in different ways. Everyone agrees that softwood pulps require refining, but they do not necessarily concur on how much or the best way to do it. Furthermore, different types of softwood pulp do require various types and intensities of refining to develop the same level of strength. Hardwood pulps may be used with little or no refining, and will generally be treated with different types of equipment. When the beater was the common refining tool, the stock could be left in it until it had received enough treatment. With continuous refiners the stock is pumped through once; if it requires more treatment, it must be passed through another refiner or slowed down on its way through the single refiner. Multiple passes can be arranged by using refiners in series, and slower flow rates can be had by splitting the flow into smaller streams and passing it through refiners hooked up in parallel.

Following the refiners, the pulps are stored to maintain a steady supply for the next stage. Stock prep operations are broken into a series of steps separated by *storage tanks*. Storage is needed to maintain the supply if mechanical problems occur and also to convert the surges of flow from a batch pulper to a continuous flow for the paper machine. The *blending and metering* step may also be operated in batches or surges. If the grade of paper being made on the paper machine is to be changed, the mixture of stock being prepared must be changed first. The change is usually made by stopping the metering and blending when it has been determined that enough stock has been prepared to finish the present order. If there is a separate blend tank, it is emptied by pumping all stock to the *machine chest*. The blend chest can then be cleaned out and refilled with the new order. When the supply in the machine chest runs out, the chest can be quickly cleaned, if necessary, and the stock for the new order pumped in, to be ready to supply the paper machine with the least time lost.

From the machine chest, the stock flows to the *paper machine*. The operations between the machine chest and the headbox may be found in the machine room and may be considered part of either the papermaking or the stock prep operations. The consistency regulator and a meter control the flow of fibers to the paper machine and help control the basis weight of the paper being produced. In the past the meter used at this point was an open box, called a stuff box, with an adjustable overflow dam and orifice. Although magnetic flowmeters are now commonly used, the name persists in some mills.

The jordan refiner used at this point in the flow is designed to favor cutting of the fibers. Cutting may not always be the most desirable form of fiber treatment, but it is the one most directly observed in the paper and most useful at this point. Cutting will have a pronounced effect on the formation or cloudiness of the paper as viewed with transmitted light. The stock flowing through the jordan is generally about 2% consistency, which is another factor that makes the jordan a cutting machine.

Fibers have a strong tendency to clump together and would make very lumpy paper if not diluted to below 1% consistency. This dilution requires a large volume of water, but large volumes of water are removed from the stock at the paper machine and can easily be pumped back to this point to dilute the stock. The screening and cleaning operations are carried out at this low consistency.

SCREENING AND CLEANING

Following dilution to below 1% consistency for the headbox, the stock is sent through *screens* and *cleaners* to remove foreign materials. The oversize materials are removed by a screen typified by the flat screen shown in Figure 5.4. The screen material shown there is a flat piece of metal with slots milled in it such that they are wider at the bottom than at the top, to help prevent plugging. The dilute stock is forced through the screen by pressure, and the screen is vibrated or the pressure is pulsed to simulate vibration. The vibration and pulsing help to prevent plugging. The slots are 0.008 to 0.010 in. wide, certainly wider than some of the fiber clumps or other material to be removed. The screen sorts the material on the statistical probability of the individual particles passing through. Small particles have a high probability and large ones have decreasing probabilities as their size increases. If the large particles are given enough opportunity to pass through, they will. Therefore, the flow rate through the screen is very important. By balancing the flow rate to the screen surface against the reject flow rate, it is possible for the slots to reject fiber bundles smaller than the width of the opening. Although the original screens were flat, operating in open boxes, higher production rates demand that we use closed screens today. Furthermore, some screens use holes rather than slots, but the same principles are used in most closed screens in today's mills. These screens remove oversize particles but will not remove heavy particles.

Figure 5.4. Section of flat screen showing screen operation

Total pulp slurry

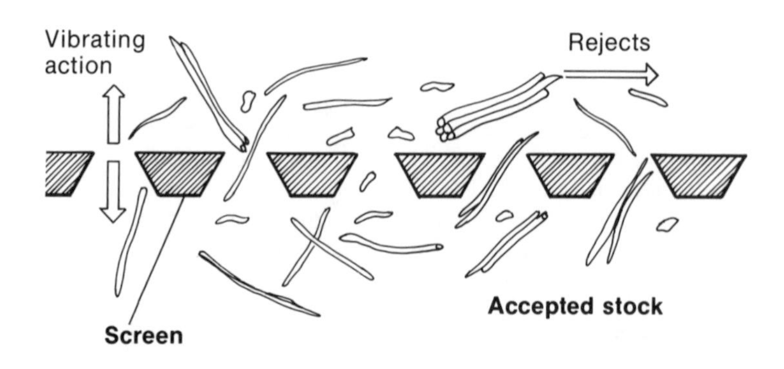

Heavy materials, or particles with a specific gravity greater than that of the fibers, are removed with a centrifugal cleaner similar to the one shown in Figure 5.5. This type of cleaner sorts material on the basis of specific gravity. The total stock flow enters the cleaner at the top left side of the drawing and is pumped into the cleaner tangent to the outside wall. This angle of entry creates a circular flow inside the cleaner. The centrifugal force created by the circular motion causes the heavier material to be thrown to the outside of the cleaner. There are also openings in the top and bottom of the cleaner; the sizes of these openings must be balanced to control the relative amounts of material that will be allowed to pass from each one. Just as in the case of the flat screen, we must throw out some good material with the bad. That is to say, in order for the heavy materials

Figure 5.5. Centrifugal cleaner

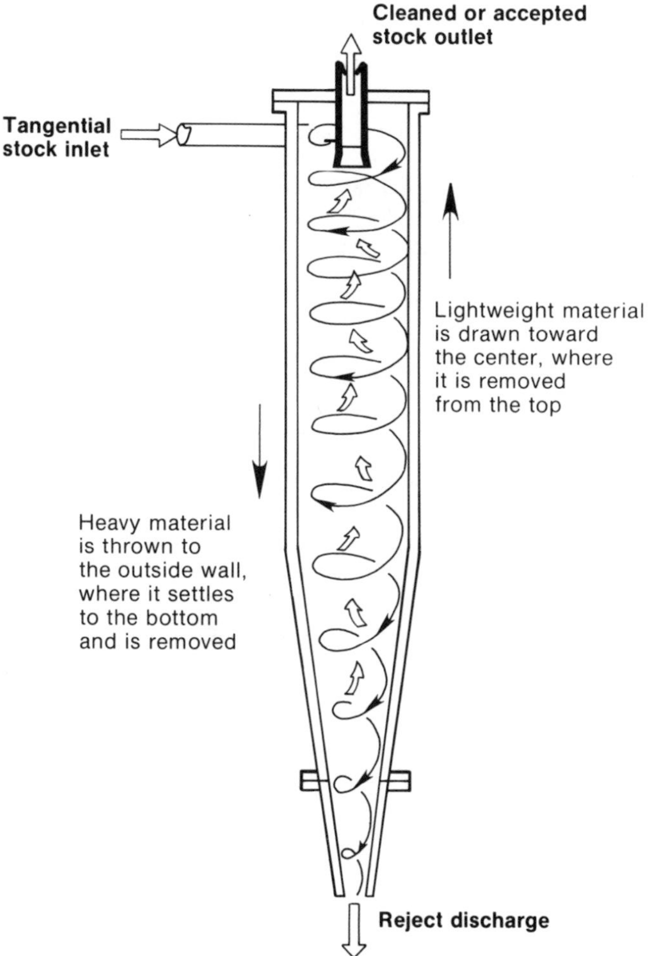

to flow to the bottom and be rejected, there must be some flow in that direction. However, if too much downward flow is allowed, enough stock will not be drawn out of the top. The balance is called the efficiency-reject ratio. If it is desired to reject all of the heavy particles (which would give us 100% efficiency), it will be necessary to reject nearly all of the material (a high reject rate). For the purpose of maintaining high production rates, it is desirable to have a low reject rate, but at the same time the efficiency of removal of harmful materials must be kept at an acceptable level. The problem is usually solved by using several cleaners in parallel to reduce the flow rate through each cleaner and also by collecting the rejected materials and recleaning them.

Since the screens and cleaners both require consistencies of less than 1.0%, they are normally installed in the line leading to the headbox. The amount of water needed to dilute the stock to this level is sizable and will not be added until necessary. For example, the water needed to dilute 2% stock from the jordan to 1% stock would be 50 tons of water per ton of dry fiber. Since the ton of dry fiber at 2% is already dispersed in 50 tons of water, that dilution would mean adding an amount of water equal to the original amount of stock. To dilute the same 2% stock to 0.5% consistency would require the addition of 150 tons of water, or a volume equal to 3 times the original. It will be seen later, in the discussion of paper machine forming devices, that enormous volumes of water must be handled to make paper. This is the reason that the stock is diluted with water from the paper machine, rather than with fresh water. There is a flow loop associated with every paper machine that removes the water in the machine, feeds it back to be used for dilution and then returns it to the paper machine.

FIBER STRUCTURE

Fibers that have been liberated by the methods described in Chapter 4 are not generally ready to be used to make paper. The notable exceptions are: groundwood pulp, which is used in newsprint, and wastepaper, used in combination boxboard for packages such as cereal boxes. Groundwood fibers are mechanically treated by the grinder and secondary fibers were already refined for their original use. If at this point in their treatment chemical-pulp fibers were formed into a pad on a screen, the dried pad of fibers would not bond together well and might even fall apart when attempts were made to remove it from the screen. The reasons for this behavior are that the fibers are relatively stiff and don't have enough bonding groups exposed on their surfaces to bond together into a strong sheet.

To better understand this behavior and the treatment required to modify it, it is desirable to study the structure and nature of the fiber. Fibers come in many sizes and shapes from different types of trees, but the predominant fibers used to develop strength in paper, technically called *longitudinal tracheids*, are from softwoods. Since these are the most important fibers for strength development, we will devote our description to them.

It is well known that wood fibers are made of cellulose. It is fairly well known also that the cellulose is somehow created as the result of photosynthetic activity in the tree leaves. Confusion and misunderstanding generally start with the fact that the fibers are actually the cells of which the tree is made and that the cellulose found in the fibers is not the immediate product of photosynthesis. Cellulose is made in the individual cells from sugars generated by photosynthesis and transported to the cells or fibers by other fibers in the tree. The cellulose molecules are formed inside the fiber and are deposited on the inside of the cell wall by the living material inside the cell. Figure 5.6 relates part of this process by demonstrating the relative size and organization of the cellulose molecules within the cell walls. Beginning in the upper part of the drawing, we see the cellulose

Figure 5.6. Relative sizes of fibers and fibrils

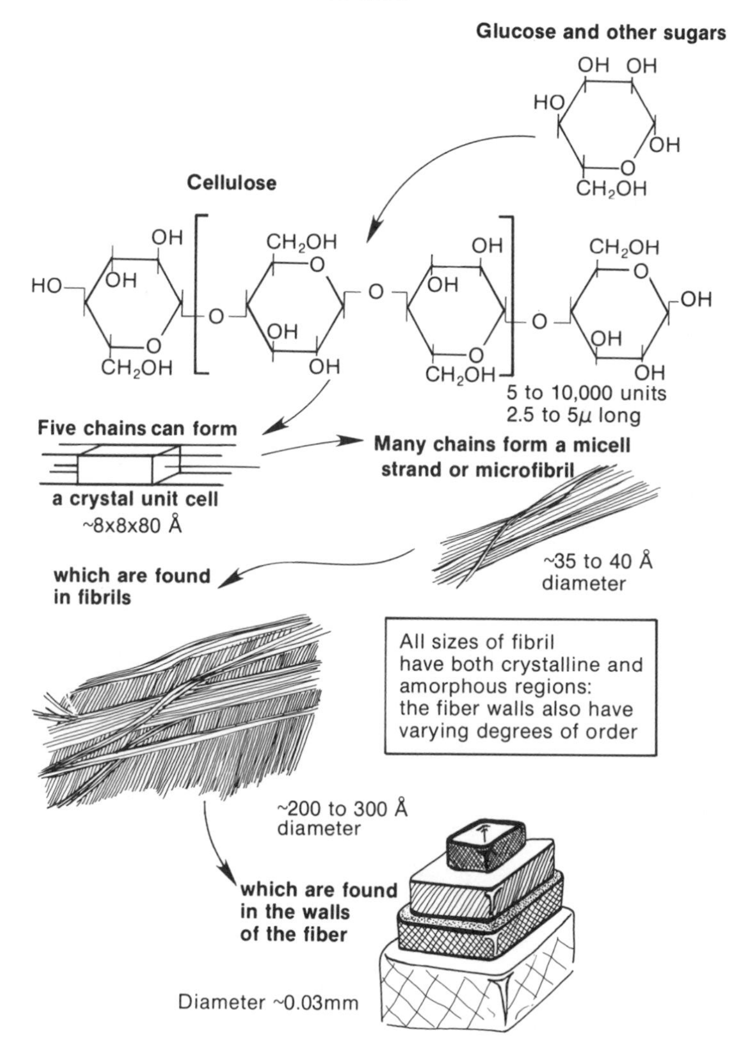

molecule itself, a ring structure of carbon atoms with hydroxyl groups (oxygen and hydrogen atoms) extending from the rings. It is these hydroxyl groups that will produce bonding in the sheet of paper. Cellulose has a tendency to form a sort of crystalline structure, but because of the size of the cellulose molecules and the other chemicals present in the fibers, it is not easy for the cell to form large crystalline areas. We find instead that the cellulose molecules pass through highly ordered or crystalline areas and then into random or amorphous areas. Figure 5.6 indicates that several cellulose chains will be loosely organized together into threads or strands, which can be found in the fiber in a variety of sizes. Within the smallest of these threads, called a *micell strand*, we find several cellulose molecules which pass through regions of high and low order. Not all of the molecules need to be included in all of the ordered regions, and some of the molecules may even extend from one strand to another, being part of perhaps several such strands. The same organization can be found for all of the sizes of elements in the fiber. The micell strands are organized into larger strands called *fibrils*, with the fibrils having ordered and random regions; and again, the strands may be common to more than one fibril. The sizes and names given to these elements in Figure 5.6 are only rough since the sizes found are not really distinct, but rather a continuous range.

Figure 5.6 gives an indication of the relative sizes of the cellulose, fibrils and fibers, but it does not indicate the true manner in which the fibrils are deposited in the cell wall. A better picture of the actual fiber structure is found in Figure 5.7, which shows a fiber situated in the wood. The fibers are held together in the

Figure 5.7. Structure of fibers

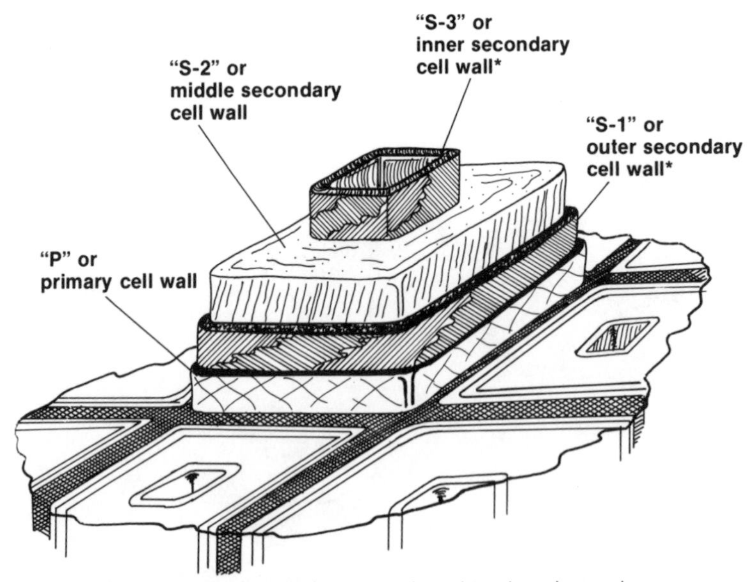

*S-3 and S-1 cell walls are drawn sectioned to show layered structure and "criss-cross" fibrillar orientation.

wood structure by a cementing substance known as *lignin*, which is found throughout the fiber, but primarily in the layers shown between the fibers in this representation. This layer, called the *true middle lamella*, will be more than 80% lignin. The fiber, when it was initially formed, had only the outer cell wall, which is called the *primary cell wall* (P). The primary cell wall is approximately 70% lignin, with the remaining amount cellulose.

The cellulosic material in the primary cell wall was presumably present at the formation of the cell by cell division. The fibrils in the primary cell wall are completely random, or nonoriented. The *secondary and inner cell walls* are all deposited by the living material in the cell. The fibrils in the outer secondary cell wall (indicated as the S-1 wall in Figure 5.7) are considered to be laid down in a sort of crisscross pattern. The first fibrils are laid down parallel to one another at an angle of approximately 45 degrees on either side of the axis of the fiber.

The thickest of the walls in the fiber is the middle secondary (or S-2) cell wall, as shown in Figure 5.7. The fibrils in this wall are again laid down in layers, but the fibrils are all oriented nearly parallel to the fiber axis. Because all fibrils are parallel, it is difficult to discern the layered structure. However, photographs of swollen or beaten fibers show breaks in the wall between layers.

The final wall deposited in the fiber is the inner secondary (or S-3) wall. Early evidence indicated that all layers in this wall were at the same angle, about 45 degrees from the axis of the fiber. However, more recent evidence indicates a structure similar to that of the S-1 wall.

The crisscross layering of the fibrils in the S-1 and S-3 walls contributes considerably to the strength of the fiber. These inner and outer walls form a sort of two-walled tube, with the bulk of the cellulosic material contained in the region between these two retaining walls. As long as these two walls remain intact, the fiber can retain its stiffness. Furthermore, the orientation of the majority of the fibrils in the fiber axis in the S-2 wall gives the fiber considerable tensile strength. The primary cell wall is of little consequence since it is removed during the pulping and bleaching operations.

BONDING AND REFINING THEORY

It is the goal of refining to break down the ordered structure of the fiber to expose more hydroxyl groups for bonding. This is accomplished by mechanical action designed to tear at the surface of the fiber and partially remove some of the fibrils, and also designed to destroy some of the order within the fiber wall to reduce the stiffness of the fiber. Again, the crisscross nature of the S-1 wall is valuable in keeping the fiber from coming unraveled. Fibrils can be torn loose from the fiber, but only in part. It is valuable for one end of the fibril to remain attached to the fiber in order to help bond fibers together.

Bonding Theory

The functions, or goals, of mechanical refining can be combined into one, that of promoting bonding in the sheet. The sheet is held together by hydrogen bonding, and the relationship between *refining* and *bonding* may be made clearer by reference to Figure 5.8. Figure 5.8 shows two fibrils with the hydroxyl groups (O-H's) aligned and bonded at the top. The important parameter that cannot be shown is the size relationship between the fiber and the hydroxyl group. The fiber may only be about 0.03mm (0.001 in.) in diameter, but the hydroxyl groups must come within about 5 angstroms (Å) of one another to bond. Although there may be a fairly high number of hydroxyls exposed on each fiber surface, the probability of large numbers of them coming into this close proximity by methods designed to randomly deposit them together is not too great. It is important to note at this point that the strength of the hydrogen bond that forms between two hydroxyls is fixed and the only way to increase the strength of the sheet of paper is to increase the number of bonds between fibers. Because of the size and spatial

Figure 5.8. Hydrogen bonding and water removal

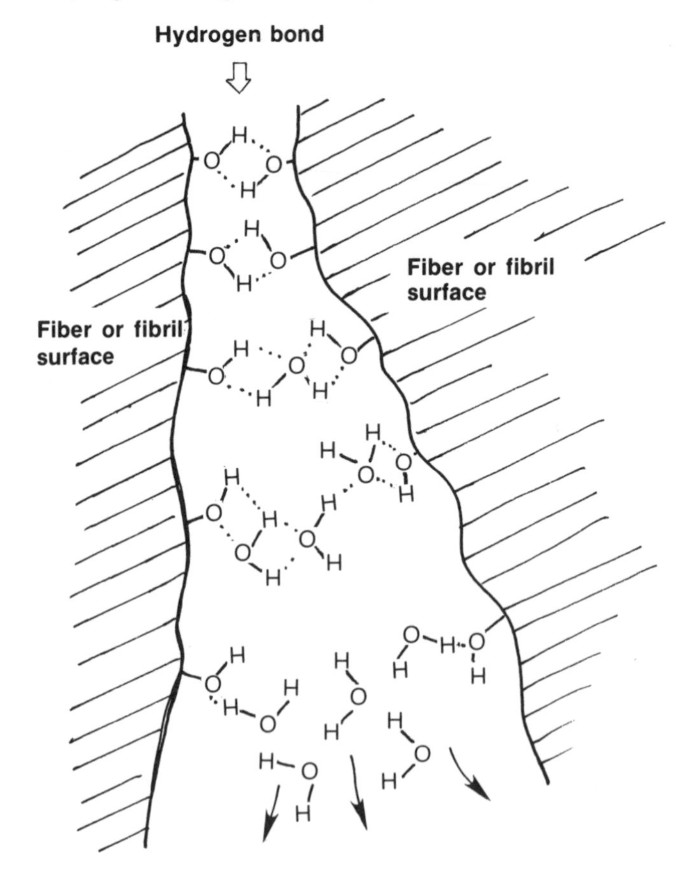

As water is removed, it helps draw hydroxyls together and align them for the development of hydrogen bonds.

problems, it is important to develop mobility of the hydroxyls, or more specifically, of the fibrils in which the hydroxyls are located.

As indicated in Figure 5.8, water plays an essential role in bringing the hydroxyls together. The polar water molecule is attracted to the hydroxyls and, as it evaporates or is forced from the sheet, draws the fibrils (or fiber parts) together and aligns the hydroxyl groups for bonding. However, unless the fiber has been treated properly, the hydroxyls cannot be moved as needed. The increased mobility comes from a combination of increased fiber flexibility, allowing the fibers to collapse when dried, and the exposing of fibrils from the fiber surface.

Fibrillation Theory

The increase in bonding area due to fiber collapse is quite obvious; the contact area between two flat ribbons is greater than the contact area between two tubes. With increased contact area there is a greater possibility of hydroxyls falling in the proper relationship to bond. Increased flexibility of the fibers can come from the removal of materials from the walls chemically during pulping, or it may come from repeated flexing of the walls during beating, which reduces the degree of order in the walls without removing anything. This increased flexibility is called *internal fibrillation*, as opposed to the actual exposing of fibrils from the fiber surface, which is called *external fibrillation.*

External fibrillation is also referred to as *brushing* and is the most important means of obtaining bonding in the sheet. The fibrils are much more flexible than fibers and can be seen in electron-microscope pictures to form a sort of "cobweb" structure surrounding fiber contact points. These cobwebs are filled with hydrogen bonds and even form newer, bigger strands or fibrils that help to hold the fibers, and therefore the sheet, together.

Internal and external fibrillation are as much theoretical concepts as they are facts. We can find evidence for the existence of both in refined fibers, so both

Light microscope photograph of paper surface, magnified about 40 times.

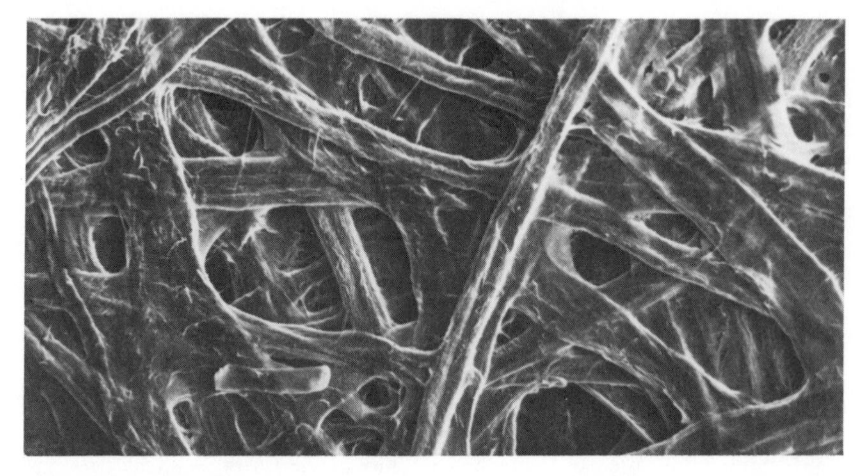

Scanning electron microscope photo of paper surface magnified about 200 times. (Courtesy Institute of Paper Chemistry)

must occur. We can even find discussions that describe types of treatment that will favor one or the other. In actuality, it is extremely difficult to obtain one form of treatment without getting some of the other.

Fiber Cutting

A third form of treatment the fibers can receive is called *cutting*. As will be shown shortly, there is some uncertainty whether the fibers are actually cut, pulled apart, or just beaten so severely that they fall into pieces. In any event, refining

Scanning electron microscope photo of paper surface magnified about 1000 times. (Courtesy Institute of Paper Chemistry)

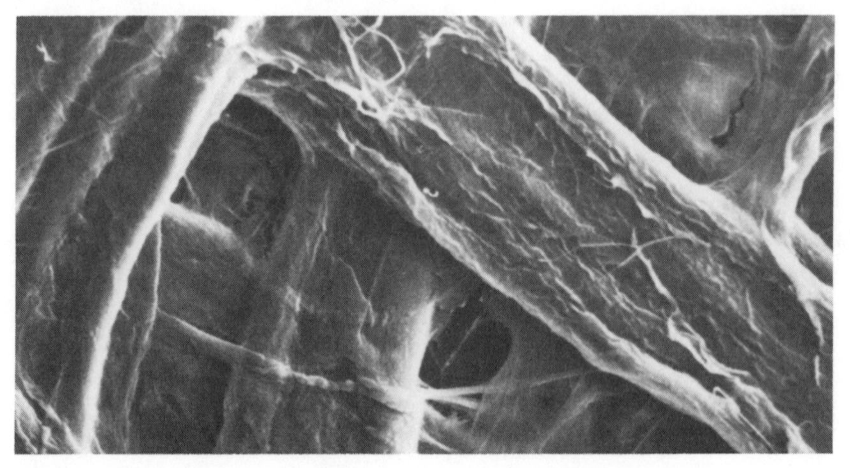

will result in a certain amount of fiber shortening. Fiber cutting has a mixed effect on sheet strength. Some shortened fibers can be shown to fill in between longer fibers and increase strength in the right circumstances. However, if the fibers all become too short, sheet strength will suffer. As with the two forms of fibrillation, it is very difficult to design a system or treatment that will cut without generating some fibrillation at the same time. However, cutting is sufficiently different to allow us to design situations that will favor either primarily cutting or almost no cutting.

One of the most important considerations in determining the relative amount of cutting is the consistency of the stock being refined. The relationship between cutting and consistency is demonstrated in Figure 5.9, which shows what may happen to fibers as two bars of a refiner pass by one another. In the upper portion of the drawing, the low-consistency condition is represented. It is possible that a single fiber could be trapped between two such bars and cut in a sort of scissors action. However, the individual fibers may be smaller than a nick or scratch in the surface of the bars. In that case it would be difficult to visualize the fibers actually being cut by such large bars. The alternate theory to explain cutting, or fiber length reduction, is represented on the right side of the upper part of the figure. It is theorized that the fibers become entangled in clumps that are too large to pass between the bars and the clumps are pulled apart by the bars. As the clumps are pulled apart, any fiber tangled in both halves of the clump will be stretched and may be broken. Either theory will suffice to explain the fiber shortening that occurs with low-consistency refining.

Figure 5.9. Effects of consistency on refining

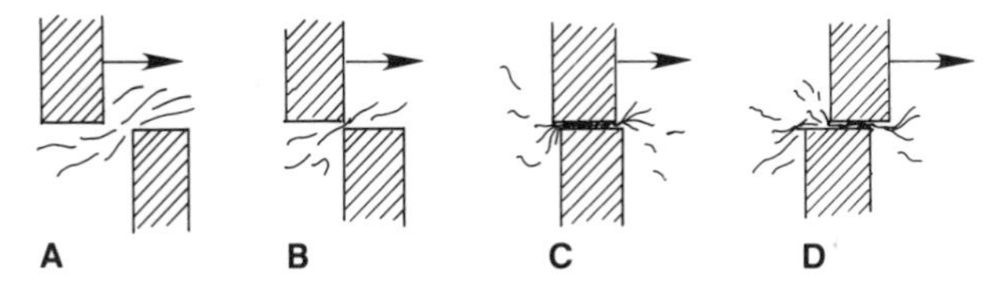

As the upper bar passes the lower, low consistency will favor either trapping single fibers and cutting them, as in B, or stretching and breaking them, as in Views C and D.

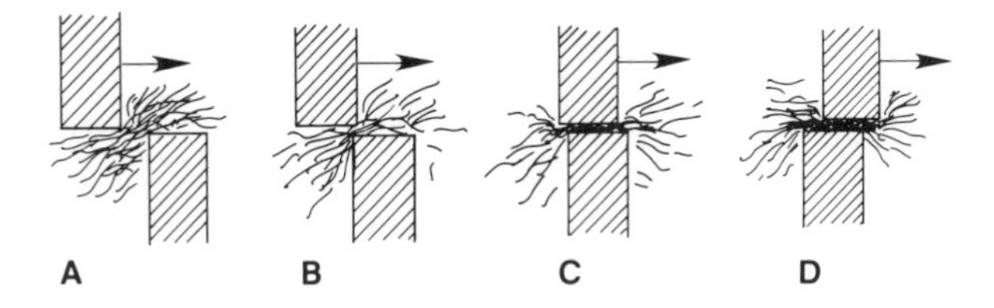

At higher consistencies, the fibers will tend to form a blanket or cushion on the bars, leading to a crushing rather than a cutting action on the fibers.

The lower part of Figure 5.9 shows the theorized blanket or cushion of fibers created between the bars at high consistencies. High consistency does not allow much length reduction, but does favor brushing and fibrillation of the fibers. The theory proposes that the energy transmitted to the fibers crushes them against one another, and therefore creates a gentler action than crushing them between steel surfaces.

It must be stressed that it is not possible to separate completely conditions that will cause only cutting from conditions that will cause only brushing. The two are opposite ends of a continuous scale. At low consistency cutting is favored, and at high consistencies of 30% or so there is almost no cutting. Other considerations, such as the hardness of the bars and the design of the refining equipment, also are important, but consistency is perhaps the most outstanding factor.

MECHANICAL REFINING EQUIPMENT

Conical Refiners

Refiners come in a wide variety of shapes and sizes. The original refiner, the beater, had a rotating wheel or roll with bars that beat against bars in a stationary plate. Although attempts were made to line up several such roll-and-plate combinations in a single beater to create a continuous operation, the beater never made the conversion to continuous operation. The closest approach is the *conical refiner* shown in Figure 5.10. The roll has been replaced by a cone with bars on its surface. The stationary plate has been replaced by a shell, or second cone, with bars on its inside surfaces. The inner cone is forced into the outer cone, causing pressure to develop between the bars of the rotor and the stator, just as the roll in the beater beat against the stator. In the conical refiner, the stock is not carried through by the rotation of the roll, but is forced through by a pump in a direction

Figure 5.10. Conical refiner

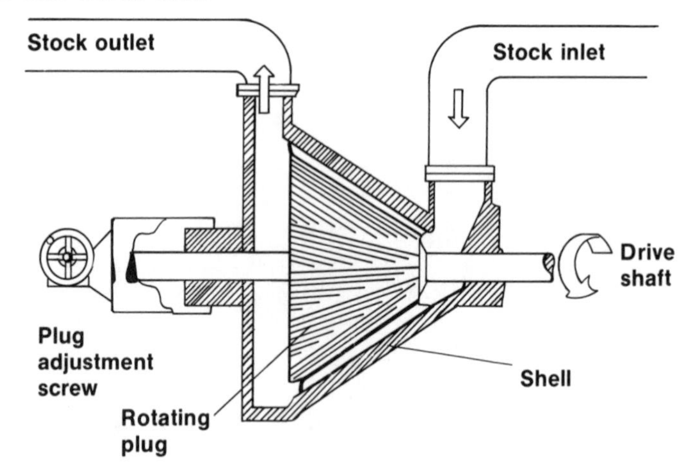

roughly parallel to the axis of the cone. Actually the combined action of the rotation and pumping causes the fibers to follow a spiral route through the refiner. The continuous refiner requires the use of a pulper to deliver defibered pulp in water suspension.

Theoretically, one could design the cone of a conical refiner to have any angle, but most are around 45 degrees. The notable exception is the jordan, which has an angle of about 12 degrees. The rotation of the plug in the refiner creates a pumping action that helps force the stock through the refiner. The magnitude of the pumping action is directly related to the speed and diameter of the cone, or the difference in diameters between the small and large ends. The resistance to the pumping action comes from both the clearance between the cone and its shell and the length of the refiner. If we fix the difference between the two diameters of the ends of the cone and change the angle, we see that the smaller angles will require a longer refiner and therefore create more back-pressure to the pumping action. In order for the stock to be pumped through, refiners with smaller angles must use lower consistency stock; the larger the angle, the higher the consistency that can be handled. Since the jordan has a small angle, it can't handle stock much above 2%; it is therefore predominantly a cutting refiner. A refiner with an angle of about 45 degrees can handle stock up to 6% to 10% consistency, and will favor fibrillation.

Disc Refiners

The *disc refiner*, in which the angle of the cone has been opened up completely, has minimum resistance to flow and handles stock up to about 30% consistency. An example of a disc refiner is represented in Figure 5.11. The disc refiner has two discs with raised bars on their facing surfaces; one of the discs rotates and the other is stationary. The stock is pumped in at the center of one of the discs and flows out between them. Some disc refiners are designed with both discs driven in opposite directions, to increase the amount of action on the fibers. These refiners

Figure 5.11. Disc refiner

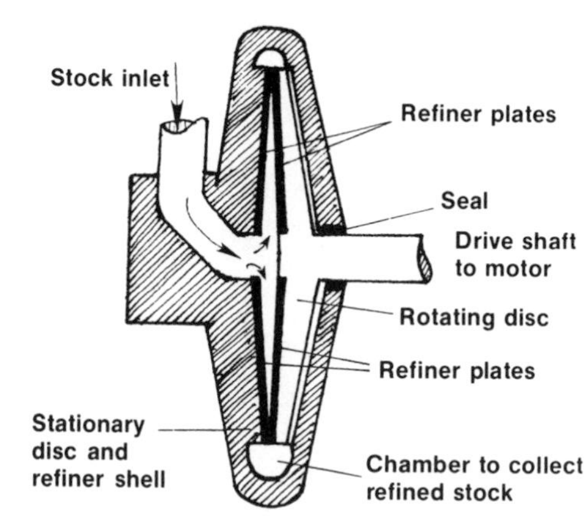

Disc refiner installed in a secondary fiber refining application.

with both plates driven are called double-disc refiners by some, but one company uses "double disc" and "Double D" as trade names for refiners in which both discs are not driven. There is also a refiner called a triple disc, which has one driven disc with bars on both sides sandwiched between two stationary discs. All of these different designs are accompanied by special claims, but in general all disc refiners favor fibrillation and do not create much cutting.

EFFECT OF REFINING ON SHEET PROPERTIES

The strength of paper is normally positively affected by increased refining. The overall strength of the sheet can be visualised as a combination of the strength of the fibers, the amount of bonding and the bond distribution. Cutting can have an adverse effect on the fiber strength, as can excessive beating. Therefore, although refining can increase the relative amount of bonding and therefore increase sheet strength, the maximum strength obtainable remains a function of the individual fiber strength and the number of fibers present in the sheet, or basis weight. The overall picture therefore is that the strength of the sheet will increase with refining until the strength of the bonds reaches the strength of the fibers. The increase in strength will reach a maximum and then decrease as the fibers are weakened. However, all of this discussion is with reference to the general strength of the sheet and is not related to any one particular strength test.

An increase in fibrillation will lead to increased bonding in the sheet. An increase in bonding will increase the fold, tensile and mullen strength since these are positively affected by bonding. The tear test, on the other hand, is commonly lowered by an increase in bonding. When a sheet is torn, some of the fibers are broken and others will be pulled from the mat of fibers with little damage. Generally speaking, as we increase bonding we reduce the probability that fibers will be able to be pulled from the mat intact. Since the internal tear test measures the work required to tear the sheet, the pulling of intact fibers from the sheet requires more work and will increase the tear value.

Another property that is decreased by beating is the opacity. The ability of a sheet to hide the printing on the back side is a function of the sheet's ability to scatter light as it passes through the sheet. As light passes through the sheet of paper, it will be scattered every time it passes from the air into a fiber, or from a fiber into air. No scattering will take place when the light passes from one fiber into another. Increased bonding will increase fiber-to-fiber contact and therefore decrease opacity. Increased beating, and subsequent bonding, will also increase the density of the paper. If the basis weight is kept at the same value and the stock is refined more, the increased bonding should lead to a lower caliper. Refining that leads to fibrillation can therefore be seen to have mixed effects on the paper. Some strength properties will be improved and other strength and optical properties will suffer. Arriving at the optimum overall combination of properties is therefore a compromise.

Refining that leads to cutting of the fibers is more complex and does not translate into individual test results as directly. In general, cutting will decrease tearing resistance the most, with the potential to decrease fold, tensile and mullen if it proceeds too far. The complex part of the relationship is in the effect of fiber length on formation. As used here, *formation* indicates the overall uniformity of the sheet with respect to fiber distribution. If the sheet is well formed, the fiber distribution is uniform and the paper, when held up to the light, will appear quite uniform. A wild formation means that the fibers are poorly distributed and the sheet is cloudy or curdled in appearance when viewed with transmitted light. In strength testing it is said that the chain is only as strong as its weakest link. The same concept applies in this situation. If a paper has a wild formation, there will be many thin places in the sheet that are weaker than the thick ones. Fiber length enters this discussion since longer fibers are more likely to clump together in the headbox on the paper machine and cause wild formation. Accordingly, if the formation is wild, and refining that reduces average fiber length is used, fiber shortening can contribute to improved formation and increased strength of the paper. If, on the other hand, the sheet is already well formed and the fibers are cut, the sheet will be weakened.

Other specific tests not mentioned are not as greatly influenced by refining as those that have been discussed. However, it must be remembered that there are many trade-offs between different desirable properties. The design and construction of a paper product involves compromises at every step to optimize the properties desired with respect to the materials available and the price that can be obtained for the finished product.

REFERENCES

Kocurek, M. J. Ed *Pulp & Paper Manufacture*, Third Ed., Vol. 6 The Joint Textbook Committee of the Paper Industry, Atlanta, TAPPI (in press).

Smook, G. A., *Handbook for Pulp & Paper Technologists*, The Joint Textbook Committee of the Paper Industry, TAPPI, Atlanta 1982.

6 Paper and Paperboard Manufacturing Operations

OVERVIEW OF MANUFACTURING MACHINERY

Before beginning the detailed descriptions of each of the parts of the paper machine, it will be helpful to review the total papermaking process. The simplest description of the total process was given in Chapter 1 as: mix, drain, press and dry. Chapters 4 and 5 were devoted to the "mix" operations needed to liberate and prepare the fibers for papermaking. The remaining three operations all take place in the paper machine, and the descriptions of the machines will be based on the same lines of separation. "Drain" becomes *sheet-forming devices.* "Press" becomes *consolidation of the web* and "dry" remains unchanged as *drying methods.* These operations are found, respectively, in the wet end, press section and dryers. The greatest diversity is in the design of the wet end, and therefore the classification of machinery is primarily on that basis.

The Fourdrinier Paper Machine

A drawing of the basic parts of a typical *fourdrinier* paper machine is presented in Figure 6.1. The machine name comes from the type of wet end used, which stems from an invention credited to the brothers Fourdrinier in about 1800. The cleaned, screened and diluted stock enters the picture in the upper left corner

Figure 6.1. Fourdrinier paper machine

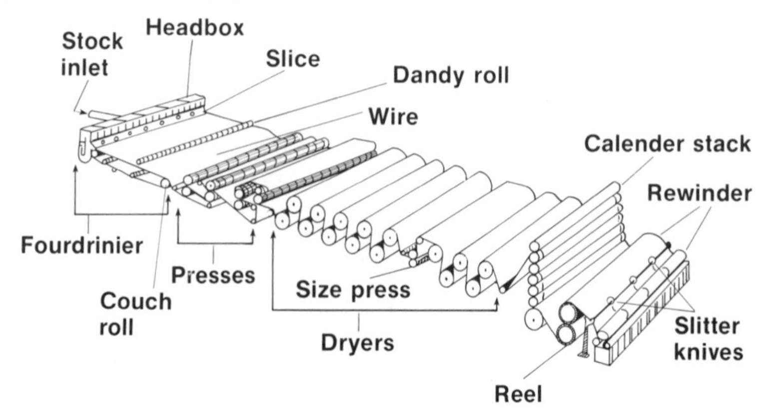

behind the *headbox.* The functions of the headbox and *slice* are to deliver a ribbon of stock to the wire at uniform dilution, thickness and speed. The headbox is a tank of the necessary size to deliver the volume of stock required and is positioned above or beside the wire so it can deliver the stock to the wire through the slice. The slice is a narrow opening in the headbox through which the stock flows. The slice must be adjustable so that the thickness of the ribbon of stock can be controlled. The headbox and slice work together to control the volume or weight of fibers as well as the flow rate (velocity of the flow relative to the speed of the wire).

The *wire* is a continuous belt of woven material, originally metal wire but now more frequently plastic materials. The wire travels over a series of rolls and/or devices called *hydrafoils,* which both keep the wire level and also remove water. As the stock and wire pass through the wet end, water is removed first by gravity, next by low pressure generated on the back side of the rolls and foils and finally by suction devices located under the wire. The paper web leaves the wire at the *couch roll* and the wire travels back below the forming table to the headbox to receive more stock and continue to form the continuous web of paper. *Showers* below the forming table clean the wire on its return to the headbox.

The web leaves the fourdrinier section at the couch roll, and enters the *press section.* The presses are hard rolls that squeeze the paper gently to remove water. Of equal importance is their function in bringing the fibers together to promote bonding. Because of this important function we discuss this section of the wet end under the heading "Consolidation of the Web."

The web leaves the press section and is passed around a series of steam-filled drums (called *dryer cans*) where the remaining water is removed by evaporation. The web travel is shown in Figure 6.1 to be such that the web passes over the top and under the bottom drum, so that first one side of the web, then the other, is heated. Although not shown in this simple diagram, *felts* are used to hold the web tightly against the drums to improve heat transfer. The felts are actually woven cloth or plastic screens that will not absorb water, but will allow the evaporating water to pass through. There are also a series of short vertical draws between the dryer cans where the web is in the open and evaporation takes place freely. Most

The wet end of a fourdrinier machine with the headbox at the right and presses and dryers to the left rear.

A closed dryer section as viewed from the winder or dry end of the machine.

machines are now constructed such that the dryer section is enclosed in a small room or hood to help prevent heat loss from the hot dryer cans. Therefore, on a visit to a papermill, the dryers are often concealed behind doors and are visible only through windows.

Figure 6.1 also shows a *size press* in the dryer section, which may be used to apply a chemical solution to the surface of the web to improve its water resistance. The chemical solutions are applied in water; therefore, further drying is required after the size press. The size press will be treated in more detail in Chapter 7.

At the end of the machine we find the *calender, reel* and *rewinder.* The calender is a stack of rolls designed to press the sheet, to smooth it and help control its thickness. The reel winds the web into a roll. The web is not fastened to the shaft of the reel, but is merely wrapped around it when the machine is started and held there by friction. If the web breaks during operation, the machine is rethreaded, and the paper is wrapped around that already on the roll and is held again by friction. When the amount of paper on the reel is at the desired level, the paper web will be broken off just before the reel and fed around the other reel shaft. Every machine has at least two winder reels for this purpose. While the paper is being wound on one reel, the other will be taken to the rewinder to be rewound.

It is not desirable to run rolls with breaks in the web through printing presses or other processes. Therefore it is necessary to rewind the paper and repair the breaks with tape or glue to give the purchaser a truly continuous web. The rewinder also must trim the web to the width needed by the processes that will follow. The rewinder therefore includes in its design *slitter knives* to cut the web into narrower webs. The knives are round discs with sharp edges and are positioned to continually cut the web as it passes under them. The rewinder obviously must be able to operate faster than the paper machine so that it will be available when the next reel is ready to be rewound.

It is easy to describe the operation of the fourdrinier machine in terms of how the paper web travels through it while it is running, but that does not satisfy the reader who wants to know how the web got there in the first place. The first operation in the startup procedure is to get all of the parts of the machine running at the same speed. With all parts running, the stock is pumped into the headbox, which has already been filled with water; water is also running through the slice and wire and back through the recirculation loop. As the consistency in the headbox increases, a web will begin to be deposited on the wire. The fourdrinier section must be equipped with a tank under the couch roll to allow the web to be dumped, mixed with water and recirculated until the thickness of the web has reached the desired level. This tank is called the *couch pit*.

There are two water jets above the wire near its edge, just before the couch roll; these squirt water down through the wet web, cutting it and removing the rough edge created by the flow of stock from the slice to the wire. The jet on the front side of the machine is fixed and the back one is movable. At startup, when the thickness of the web is proper, the movable back squirt is pulled toward the front of the machine until it is just a couple of inches from the front squirt. The narrow tail of paper between the two squirts is removed from the wire and fed into the press section. This tail can then be widened to a more manageable width, perhaps a foot, and the wider tail fed through the rest of the machine. As the tail successfully passes from the press section, the squirt is pushed toward the back side of the machine, allowing the web to widen to the desired width. All of the wastepaper, called *broke*, that is made during startup and that results from any breaks that might occur later is collected and can be reprocessed into paper.

Materials Balance

The other basic consideration essential to an understanding of the overall papermaking operation is the materials balance. Figure 6.2 gives a simplified balance for a fourdrinier type machine. The basis for the numbers given is 1 ton of fibers being made into paper. As shown in the upper right in the drawing, paper is

Figure 6.2. Paper machine material balance

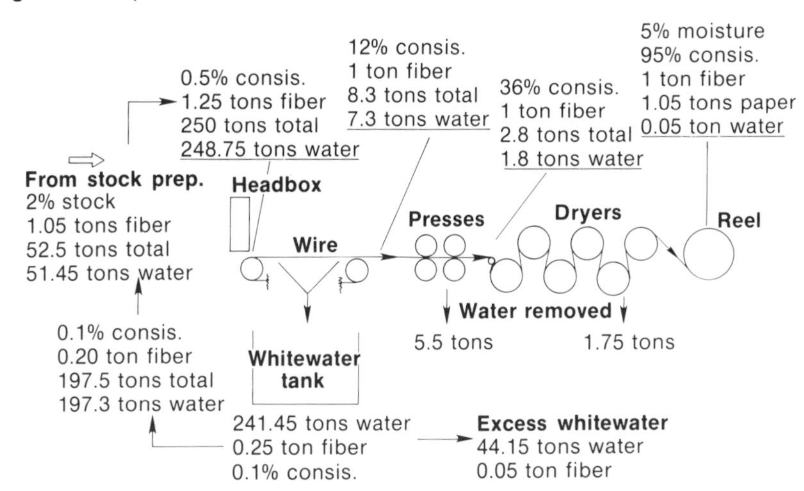

about 5% moisture under normal conditions. Therefore, 1 ton of fibers yields about 1.05 tons of paper. If we start the analysis as the stock comes from the headbox, we see that we have 1.25 tons of fiber. The stock flowing onto the wire is separated into fiber and water by the wire, but the separation is not perfect. Some of the water leaves with the fibers and some of the fibers pass through the wire with the water. The term used to represent the percentage of fibers that remain on the wire is *retention.* The calculations in this example are based on a *one-pass retention* of 80%. It is called a one-pass retention because it only considers this small section of the total machine. We will develop an overall retention figure later. Assuming that the headbox consistency will be 0.5%, which is a reasonable average, we see that the total flow through the headbox will be 250 tons of stock, which contains 1.25 tons of fiber and 248.75 tons of water.

Water can drain freely from the fibers on the wire, but without the aid of suction equipment the consistency will not get much above 10%. With suction equipment consistencies of about 16% are possible. If we assume a consistency of 12% for the sheet leaving the wire, we see that the ton of fibers will take with it 7.3 tons of water, indicating that 241.45 tons of water passed through the wire. Since we also lost 0.25 ton of fiber through the wire, the whitewater collected under the wire will average out to 0.1% consistency. The water that comes out closest to the breast roll will be richer in fibers than the water removed further down the wire or through the suction devices. The richest water is usually collected separately and sent to the stock prep area to be used to dilute the fibers coming from there.

The actual amount of stock needed from the stock prep area and the amount of whitewater to be mixed with it must be calculated together using simultaneous equations, or simply arrived at by trial and error, as has been done on machines for years. For the sake of this example, we can see that if we assume the incoming stock is at 2% consistency, and we assume that all of the whitewater is used in dilution, we need 1.05 tons of fiber from the stock prep area. This figure is used to calculate the *overall retention.* If we deliver 1.05 tons of fiber to the paper machine and get 1.0 ton of fiber out in the paper, the overall retention is 95.2%.

Since the stock coming to the wet end is at a higher consistency than the web leaving it, we have a net gain of water in this part of the machine. This net gain of 44.15 tons of excess whitewater also contains 0.05 ton of fiber. This may not sound like much, but since it is only based on 1 ton of fiber, it must be seen as a loss of 5%. If the machine was making 100 tons per day (tpd), the loss would be 5 tpd. It therefore is necessary for the mill either to send this flow to a device called a *saveall,* which will reclaim the fibers and clean water for reuse or to use this water for other dilutions where the fibers will not be lost. The latter is the cheapest and easiest way to reclaim the water and fibers, but there may be chemical concentrations in the whitewater that will not allow it to be used in dilution of the incoming stock, or for other uses.

The management of the whitewater balance on the wet end of the machine is extremely important to the economic success of the mill. The material balance also emphasizes the heavy dependence of the industry on an adequate water supply. The location selected for a mill must have water. Fibers can be shipped in from other locations, but there must be water at the mill site. Even with the large amount of water needed in the wet end of the machine, the total consumption of the mill will be well below that level, due to recirculation within the mill. The actual amount of water required will depend on how well the water is reused in the mill and what other operations are performed there, such as pulping, bleaching and coating. Based on 1972 average figures, a mill making 100 tpd of paper would be circulating about 6 million gallons per day (gpd) through the headbox, requiring about 2 million gpd in fresh water and sending a similar volume to the waste treatment plant. Since those figures were published, many mills have eliminated their effluent and are not sending any water to waste treatment. Because some water is evaporated and some is sent out the door with the paper, there will always be a need for some fresh water to be used by the mill.

The water used in the papermaking process becomes high in biological oxygen demand (BOD) and suspended solids and must be treated before discharge into a receiving stream (rivers, lakes, and so forth). Water quality regulations restrict the pollutants that a mill can discharge. Since raw, untreated wastewater from a mill normally far exceeds permitted levels of pollution, as much as 90% of the pollutants may have to be removed before discharging. Naturally, the first step in controlling effluent discharge is to reuse as much of the whitewater as possible. However, there are limitations on internal reuse, so most mills have extensive primary and secondary wastewater treatment plants. Primary treatment consists of removing suspended solids (fibers, clays, and so forth) by settling in a clarifier. Secondary treatment to remove organic materials, or BOD, is normally done by treating the effluent with oxygen in large basins. The oxygen, along with bacteria, oxidizes the organic materials, thus lowering the BOD. There are many different processes suitable for secondary treatment, and a mill must choose the one most suited to its location, costs and treatment requirements.

Returning to the paper machine balance in Figure 6.2, we see that about 5.5 tons of water are removed in the press section. This water may be collected and reused, or it may contain chemicals or other contaminants that prevent its reuse. The water lost in the dryers is indicated to be 1.75 tons per ton of fiber. The average figure quoted for loss of water in the dryers is about 2 tons of water per ton of paper.

The pulp and paper industry is highly energy intensive, ranking third in the USA after primary metals and chemicals in purchased energy consumption and accounting for about 3% of total U.S. energy consumption. An average of 30 million Btu's are required to manufacture a ton of pulp and paper; about 40% is required in the chipping and pulping operations, another 40% in drying and finishing and the remaining 20% in bleaching, washing and refining.

The industry is unique in that a significant portion of the total energy required is self-generated from fuels such as spent pulping liquors and woodwaste. As a result, pulp and paper manufacturers have been able to make significant reductions in their requirements for fossil fuels and other forms of purchased energy. In the early 1970s the industry purchased approximately 60% of its energy from sources outside the mill. This share of power from outside sources has now been reduced to close to 40% (Table 6.1). Under the pressure of energy price increases, the industry expects to further decrease its reliance on purchased energy by using more self-generated and waste fuels and process efficiencies.

Table 6.1. U.S. Pulp and Paper Industry Estimated Fuel and Energy Use, 1972–1988

	1988		1972	
Source	**Consumption (trillion Btus)**	**% of total**	**Consumption (trillion Btus)**	**% of total**
Total purchased fossil fuel and electricty	1,011	43.5%	1,243	59.7%
Purchased electricty	161	6.6	93	4.4
Purchased steam	21	0.9	23	1.1
Coal	336	13.9	225	10.7
Residual fuel oil	182	7.5	447	21.2
Distillate fuel oil	10	0.4	22	1.1
Liquid propane gas	3	0.1	3	0.1
Natural gas	338	14.0	443	21.0
Other purchased energy	2	0.1	2	0.1
Energy sold	(40)	—	(13)	—
Total self-generated and waste fuels	1,360	56.5	845	40.3
Hogged fuel, 50% moisture	267	11.1	42	2.0
Bark, 50% moisture	126	5.2	95	4.5
Spent liquor, solids	944	39.2	700	33.2
Self-generated hydroelectric	12	0.5	9	0.4
Other self-generated energy	12	0.5	3	0.2
Total energy	2,373	100.0	2,093	100.0

Note: Columns may not add to total due to rounding.
Source: American Paper Institute, as given in *Pulp & Paper 1989 North American Factbook*, p. 64.

The Cylinder Machine or Paperboard Machine

The fourdrinier machine just described is excellent for the production of lightweight paper. However, manufacture of heavy paper or paperboard on the fourdrinier machine requires the delivery of a large amount of stock to the wire, which drains slowly and requires that the machine be run at reduced speeds. The solution to this problem is found in the use of the *multicylinder machine*. Figure 6.3 presents a drawing of the wet end of such a machine. The paper is formed on the surface

Figure 6.3. Cylinder board machine

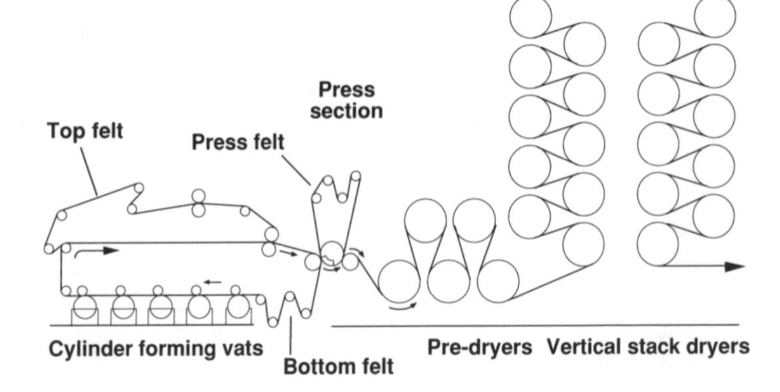

of wire-covered cylinders (shown as circles) in the cylinder forming vats. The vats are partially filled with dilute fiber suspensions similar to the stock pumped to the headbox of the fourdrinier. As the *forming cylinder* rotates through the vat of stock, the water flows through the wire on the surface of the cylinder and the fibers catch on the wire, forming the web of paper.

A *felt* (actually plastic threads woven into a continuous belt; labeled "bottom felt" in Figure 6.3) is pressed against the wet web by the *couch roll* at the top of each forming cylinder. The wet web will stick to the wet felt and be removed from the forming cylinder at this point. The felt and web picked up from the first cylinder proceed to the second cylinder, where they are pressed against the web formed there to pick it up. The felt continues through the forming section, picking up sheets from all of the vats in succession. If each cylinder forms a web that is 0.003 in. thick, a machine with five cylinders can produce a web with a total thickness of 0.015 in. Since the web is formed in thin sections, the machine can be run faster than if the total thickness were made in one layer. Another advantage of this type of former is that the uniformity of each layer will be better than if the total thickness were made in one former.

Perhaps the greatest advantage to the use of this type of former is that the different plies do not need to be made from the same stock. It is customary to make board that is white on the outside and gray on the inside. The gray inside layers are made from a mixture of old newspapers, corrugating and carton clippings. The wastepaper is fed through a pulper to break it down into fibers; the dirt is removed in screens and cleaners; then it is diluted and pumped to the machine. This type of raw material makes a gray-colored web called *chipstock*. The white layers of the web may contain bleached kraft pulp or white wastepaper stock. The use of the white stock on the outside gives the paperboard a better appearance and a better printing surface. The use of different types of stock gives the board the name *combination boxboard*. The combination of good printing quality, low cost and good stiffness makes this material an excellent packaging board and it is widely used for cereal boxes and other packaging applications.

After all of the layers are picked up by the felt, the web is sandwiched between two felts and pressed to remove the water and promote bonding. Since the individual webs are pressed together at consistencies of about 10% to 12%, they will bond together. Bonding can become a problem if the consistencies are not carefully controlled and if the compositions of the individual layers are not kept similar in certain respects. Bonding problems are seen as delamination of the paperboard during use and referred to as *plybonding problems*.

The main objectives of pressing are quite similar for either type of machine; the removal of water and condensation of the web. However, the added thickness of the paperboard web requires some machinery differences. Paperboard needs to be pressed more gently in the initial stages, and may require more nips to remove the necessary water. The paperboard machine will likely use more double felted nips, softer press rolls and another newer device called the extended nip press. This press replaces one of the rolls with a belt, which is pressed against the felt and web to allow a longer dwell time for the pressure in the nip.

The greater thickness of the web requires that it be passed through more dryers than the web in the fourdrinier machine. Because of the larger number of dryer cans, the arrangement will sometimes be vertical stacking, as shown in Figure 6.3. If the cans were stacked two-high as in the fourdrinier machine, the machine would be too long. The combination boxboard machine generally will not have a size press. Any chemicals to be applied to the web are applied in special devices on the calender stack or in conversion operations off the machine.

Since there are no squirts on the cylinders, the operators cannot make a tail there to thread the machine. Accordingly, the startup procedure will be different. The web will be started in the cylinders similarly to the fourdrinier, but the total web will be carried through the press section, where a tail will be cut on one edge with a knife; this tail is then threaded through the rest of the machine. Once the tail has been started, the man holding the knife carries it across the machine, cutting the tail out to the full width of the machine as he goes.

The cylinder board machine must also be equipped with a *winder* or reel, as was described for the fourdrinier machine. Many of the older cylinder board machines were equipped with sheeters on the dry end to cut the web immediately into sheets. Heavy or thick paperboard is rather stiff, and if it is wound onto a reel and kept there for any length of time, it may become curled, a condition called *roll set*, making it difficult to process later. Therefore it is better to sheet it immediately or wind it on large-diameter cores. Newer printing, diecutting and cartonmaking lines are being developed that use rolls rather than sheets. Therefore, most machines today wind the web on reels and are equipped with rewinders that will cut the web into narrower rolls, much the same as on the fourdrinier machine.

A water balance similar to the one shown in Figure 6.2 could also be developed for the cylinder machine. The flow for this type of machine is more complicated since there are several forming devices operating at the same time and the water from vats using the same stock (chipstock) may be blended together. Similar consistencies will be found in the sections of the machines.

FORMING DEVICES (THE WET END)

The basic function to be performed by the wet end is to separate the fibers and water in such a way as to form the web of paper. Execution of this simple objective becomes complicated by our desire to have the machine produce a uniform web on a continuous basis and with uniform specific properties. The uniformity is the main problem and affects many facets of the designs. Uniformity of thickness or bulk of the web on a microscopic scale requires that the fibers be diluted and mixed to prevent their natural tendency to *flock* (clump together) and make lumpy paper. Uniformity of the web from side to side on the machine places many demands on the design of the system that delivers the dilute stock to the wire. All of these demands, and others that have not been mentioned here, have led to the development of many different headbox and approach-flow-system designs.

In addition to the initial concern of delivering the stock to the wire properly, the demand for different paper properties has led to the development of many different devices to produce the web of paper or paperboard. Even within the fourdrinier and multicylinder categories just discussed, there are many differences that make each machine unique. Every machine is different since it is custom built to produce certain grades. Furthermore, after the machine has been in operation for any length of time, the mill will tend to change and modify it, making the differences between machines greater. It therefore becomes impossible to describe an "average" machine. Even if one specific grade of paper is selected, it is impossible to arrive at an average.

Preforming Conditions—The Approach Flow

Once the stock has been prepared in the beater room, a series of operations is needed to deliver it to the headbox and ultimately to the wire. The flow needs to be converted from a pipe to the slice in a uniform manner; chemicals and fillers need to be added and mixed uniformly; impurities and entrained air need to be removed and the flows and consistencies carefully controlled. The screens and cleaners described in the previous chapter are typically used. Special care should be exercised in the design of tanks and piping flows to ensure adequate and thorough mixing of all additives and to eliminate the inclusion of air. The fan pump is an excellent mixer and can be used for the additives; but if air is sucked into the inlet, it can result in foam in the headbox and pinholes in the paper. Air can be removed from the flow ahead of the headbox by the use of a Deculator or vacuum-deaeration device. If several additives are needed and all cannot be added at the same time, static mixing elements may be used to ensure uniform dosage of the stock.

To minimize variations in web properties across the machine, it is necessary first to convert the flow to the forming device from a pipe to a wide flow, which will eventually pass through the slice as a narrow ribbon of stock. Although this may sound like a simple task, it is also necessary that the flow rate, consistency and degree of turbulence of the ribbon all be uniform across the machine. Some early designs are shown in Figures 6.4 and 6.5. The early machines were rather simple and had few parts, not having the degree of specialization of today's machines. The *cylinder former* (Figure 6.4) is an excellent example of this simplicity. The stock enters from a single pipe in the lower left side of the vat and the flow is spread across the width of the machine simply by running into and being forced to flow over a simple dam between the pipe and the cylinder. At the slow flow rates of these older machines, this simple approach to the problem was fairly satisfactory.

The *early fourdrinier* headbox shown in Figure 6.5 used dams and other such flow-spreading devices in the box, much like the cylinder machine. The addition of the header pipe in the incoming flow stream allowed the flow to be separated and introduced to the box at three places. This modification helped spread the flow somewhat and boxes like this could be used for machine speeds up to 400 fpm. Other forms of this same idea have been used where the pipes are split again and again in a sort of tree or candlestick effect. Although helpful, this modification was not completely successful and designs of this type are rare.

Figure 6.6 shows the *tapered header*, which has become the most common solution to the flow-spreading problem for intermediate speed machines. The stock is introduced at the wide end of the manifold; as it is forced to flow out the side through the tubes into the headbox, the taper maintains the pressure and degree of turbulence in the flow. Turbulence in the flow is important, and another reason for rejection of the other designs. The corners of the square-shaped boxes provide areas in which the flow can slow down or become stagnant, allowing the fibers to clump together and perhaps even allowing slime or other bacteria to grow. Figure 6.6 shows the inclusion of a perforated hollow roll, called a *rectifier roll* or *holey roll*. This roll is rotated in use, causing the stock to flow in through holes in the one side, be mixed with other flows, and be passed out the holes on the other side. These rolls can help to even out the flow rate across the machine and also help keep the fibers from flocculating and causing lumpy paper.

Headbox designs at this time seem to be developing in two different directions. One direction is the fairly large *pressurized headbox* (similar to the one shown in Figure 6.7), which uses rectifier rolls to generate uniform flow and turbulence to prevent fiber flocculation. The other direction is toward the use of the *bunched-tube* design. Figure 6.8 shows the use of the tapered manifold inlet and the back-pressure generated by the small tubes to give uniform flow across the face of the machine. The narrow tubes also generate turbulence, which helps prevent fiber flocculation. The headbox in Figure 6.9 uses long, thin sections to control flow rates and generate turbulence to control flocculation. Both the rectifier-roll and the bunched-tube/high-turbulence designs are being used successfully. The rectifier roll headbox is more commonly found on the fourdriniers of the type described so far in this book, called *flat fourdriniers*. The bunched tube or high-turbulence type is smaller and has found wide acceptance in cylinder formers or *twin-wire machines*. The twin-wire machines use two forming wires with the headbox squirting the stock between the two and the water being removed from both directions, through both wires. The choice of a headbox is determined by the space available and the desire of the purchaser rather than firm evidence of the superiority of one over the other. Comparisons of the two designs may show some

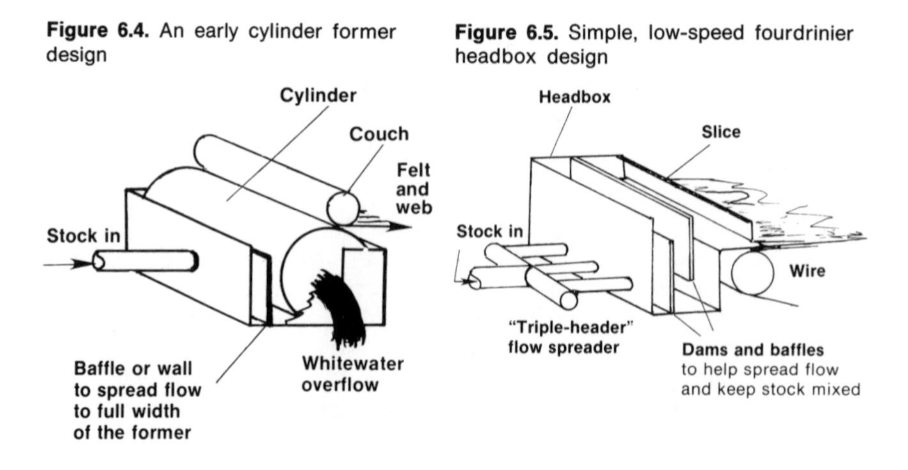

Figure 6.4. An early cylinder former design

Figure 6.5. Simple, low-speed fourdrinier headbox design

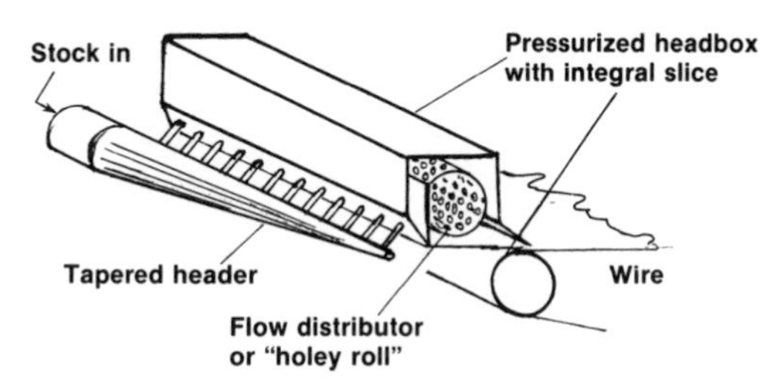

Figure 6.6. Rectifier roll headbox with tapered manifold inlet

differences, but these usually are either overcome by other aspects of the overall machine design or too small to have an appreciable effect on the paper properties (Kallmes 1979).

As machine speeds have increased, so has the potential for hydraulic pulsations to be carried through this last type of headbox. The pulsations can originate from the fan pump, rotary screens and even poorly designed piping systems. The different sources of pulsations have characteristic frequencies, which can help in their detection and identification. Because of this factor, the headbox designed with an air pad to dampen these pulsations has a slight advantage. For headbox designs that do not have the air pad, it is possible to

Figure 6.7. Typical rectifier roll headbox

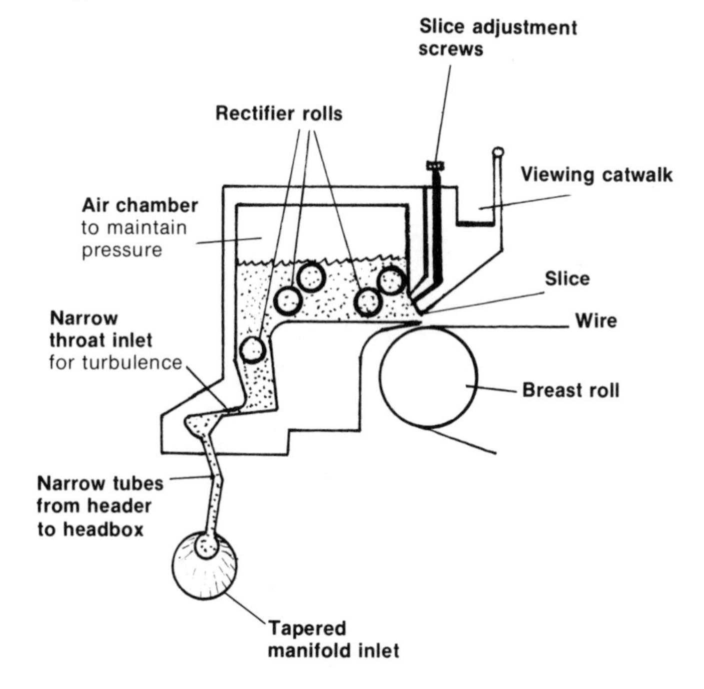

incorporate a special tank designed to eliminate these pulsations (Heissenberger, 1986 Papermakers Conference, 273–78). These tanks can incorporate drilled diffusor plates and pipes and an air pressure pad, as well as a design that changes the flow in direction and shape from the vertical pipe to the horizontal rectangular.

Delivery to the Wire (Headbox and Slice)

The slice is intended to control the flow of stock onto the wire. It cannot function alone, however, but must work in conjunction with the headbox. If we want to increase the flow rate from the headbox, we can either increase the head (the height of liquid) in the box or make the slice smaller. Since the size of the slice opening also affects the basis weight of the web, we normally increase flow rate by increasing the head in the headbox. If we refer back to Figure 6.5, it can be seen that the headbox could be made taller to hold more liquid. This solution to the problem has been used and is acceptable for speeds up to perhaps 700 fpm. As the height of the water increases, so does the probability of stock flocculation due to slow flow in some parts of the box. Accordingly, headbox design for high speeds has gone to the totally enclosed types shown in Figures 6.6 through 6.9. The enclosed headbox can be pressurized to create the desired pressure or head without having a large volume of stock in the box.

The flow rate from the box relative to the speed of the wire is important for controlling the properties of the paper. It is easy to understand that if the stock is

Figure 6.8. Static-element, turbulent-flow headbox

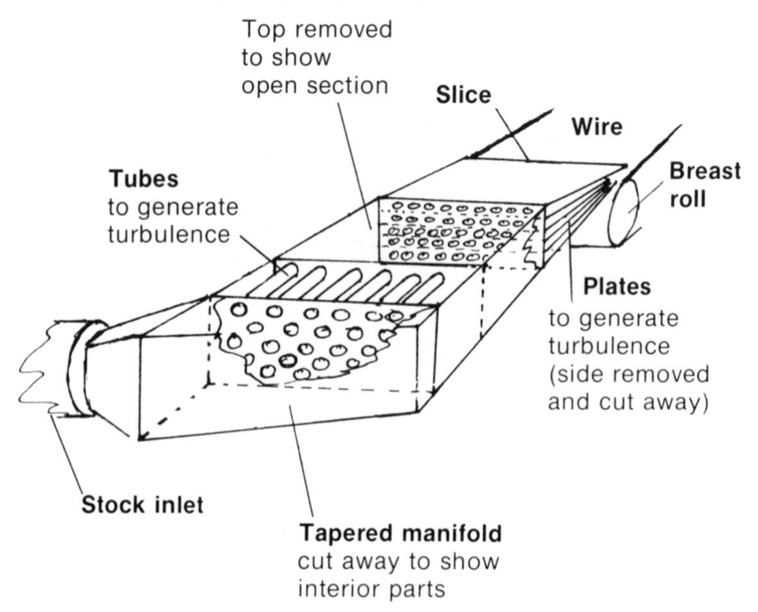

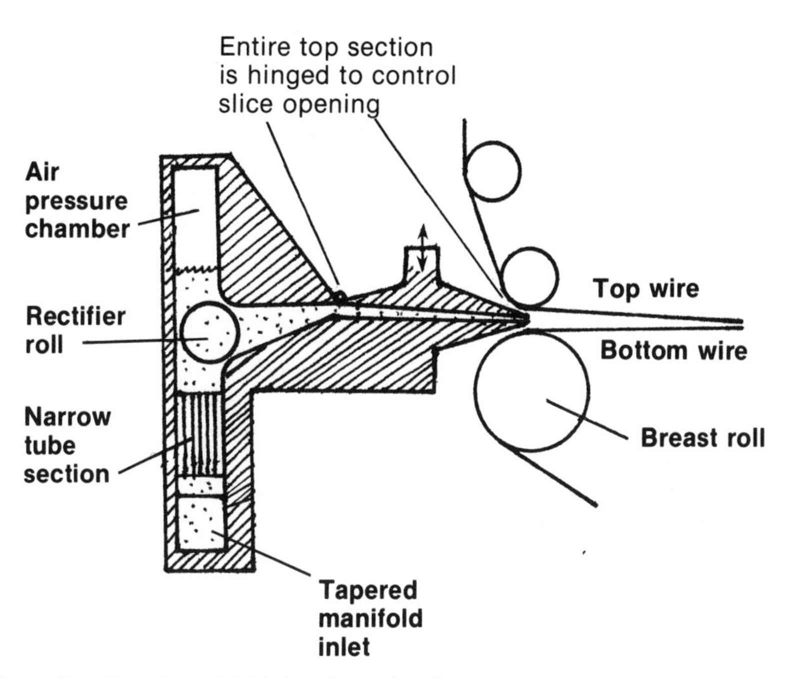

Figure 6.9. Thin-channel, high-turbulence headbox

coming from the slice more slowly than the wire is moving, it is possible for the fibers to be caught by the wire and lined up in the machine direction. If the flow from the slice is too fast, the fibers will tumble over one another or even form a permanent wave pattern in the paper like waves breaking on the beach. This flow rate from the slice is called the *spouting velocity* because on high-speed machines the stock actually spouts or is squirted for a short distance from the slice before it contacts the wire. The ratio of the two speeds is called the *rush-drag ratio*, and although important, takes us beyond the scope of an elementary treatment of the subject.

The slice also serves to control the basis weight or thickness of the web. By controlling the thickness and flow rate of stock to the wire while keeping the consistency of the stock constant, the weight of the web is controlled. A sample calculation to demonstrate these relationships is shown in Figure 6.10. By using the slice opening, machine width and machine speed, we can calculate the volume and then the weight of dilute stock delivered to the wire every minute. By multiplying the weight of the stock by the consistency of that stock, we can determine the weight of fibers delivered to the wire every minute. The machine speed and width combine to tell us the area over which these fibers will be spread. We then only have to determine the ream size we wish to calculate, and the number of those reams made every minute is easily found. Dividing the weight of the stock by the number of reams gives us the weight per ream, or basis weight, of the web being produced.

The size of the slice opening must be controlled by two different sets of controls. The total slice must be able to be moved when changes in basis weight or speed of the machine require that the total web be changed. The slice also needs a set of profile adjusters to allow individual adjustment every several inches across the width of the machine. These profiling adjusters can be used to compensate for slight differences in flow rate coming to the slice or other vagaries of the total headbox design. These adjustments will help the operator produce a web that is more uniform in basis weight and caliper across the width of the web.

The Wire

The fourdrinier and cylinder wires are similar and are a woven material originally made of brass wire. The brass resisted chemicals, but did not resist wear too well, and has been replaced by stainless steel in some applications and plastic strands in others. Although a great many varieties of weave designs are possible and a number of mesh sizes are used, the wire is commonly about a 100-mesh weave. One hundred mesh means that 100 wires/in. are used in the weaving of the screen or wire. If the wire were 100 by 60, it would mean that there are 100 wires or strands in one direction and 60 in the other. The actual size of the holes

Figure 6.10. Factors determining basis weight

Slice opening	0.5	in.
× Machine width	× 200	in.
= Slice opening area	100	$in.^2$
Slice area	0.69	ft^2
× Machine speed	× 1,000	ft/min
= Volume produced per min	694	ft^3/min
× Weight of stock per ft^3	× 62.4	lb/ft^3
= Weight of stock delivered to wire per min	43,305.6	lb/min
× Consistency of stock	× 0.005	$\frac{\text{lb fiber}}{\text{lb stock}}$
= Pounds of fiber delivered	216.5	lb fiber/min
Machine width	200	in.
× Machine speed	× 12,000	in./min
= Area of paper made per min	2,400,000	$in.^2$/min
÷ Ream area (25 × 38 × 500)	÷ 775,000	$in.^2$
= Reams made per min	5.05	

Divide 216.5 lb fiber/min by 5.05 reams/min and the basis weight of 43 lb/ream is obtained.

is a function of the mesh size, the diameters of the strands used and the style of weave. The combination of all these factors gives the papermaker a wide selection of wires from which to choose.

The selection of the best wire for each machine is based on many factors including: the amount of wear expected from friction with nonmoving parts; the chemical resistance needed due to chemicals in the whitewater; the flexibility of the fabric and how well it will stay on the machine. The wire is definitely not a permanent part of the machine; it is expected to wear and to have to be replaced periodically. Depending on the machine and the grades produced, wires may last only a matter of weeks or months on fourdrinier machines but may last for a year or more on cylinder machines. The fact that the wires must be replaced means the machine must be designed to allow their replacement. The wire is woven or welded together to be delivered to the papermill as an endless belt, and must be installed in that way. If there were a noticeable seam in the wire where it had been fastened together after being installed, that seam would mark the web on every revolution and ruin the paper it touched. Fourdrinier wet ends therefore are designed to allow removal of one side so that the wire can be slipped around all of the rolls of the wet end. Replacement requires considerable downtime for the machine and therefore lost production. The wire itself is expensive and the cost of installation and lost time add to that expense. Therefore the mill will attempt to find a wire that will last as long as possible.

The design of the wire is also important to the quality of the paper or paperboard. If the wire is too coarse, the high points of one strand passing over another, called *knuckles*, can leave marks called a *wire pattern* in the surface of the web. Early forming wires were woven with a simple over-and-under type weave of as small a wire or thread as possible to minimize pattern transfer to the web. However, if the weave is too fine, it can interfere with the removal of water from the web, or wear out too rapidly. Newer wires are complex multilayered structures intended to maximize drainage and wear and minimize marking of the web.

Fourdrinier Design

A *flat fourdrinier* design that contains most of the elements common to this type is found in Figure 6.11. The stock is delivered to the wire at the left side of the drawing, just beyond the top of the breast roll. The first part of the flat fourdrinier contributes significantly to the quality of formation of the web. The fibers can reflocculate during this portion of the operation until enough water has been removed to freeze the formation. The flat fourdrinier is usually built so that the wire can be vibrated from side to side, called the *shake*. The shake and microturbulence created on top of the foils or table rolls are important in controlling formation in this section. The exact point at which the stock meets the wire varies. If the stock is dumped on top of the breast roll, the suction on the back side of the roll will remove water quickly. If the headbox is bought close to the wire so that the bottom lip of the slice is practically in contact with the wire and the top lip extends past the bottom lip, the stock is forced through the wire in a manner referred to as *pressure formation*. Quick water removal is desirable if it

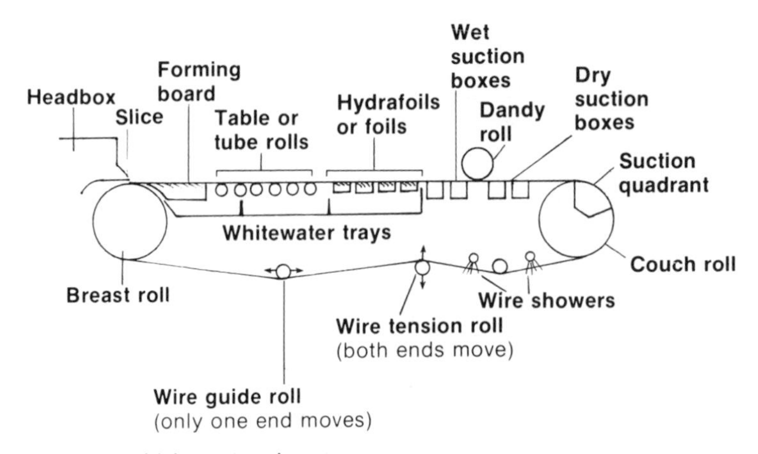

Figure 6.11. Fourdrinier wet-end parts

freezes good formation from the headbox and undesirable if it removes too many fines and fillers from the web. The rest of the water removal equipment is discussed in the next section.

As the web leaves the wire at the couch, the wire returns to the breast roll to receive more stock and continue formation of the continuous web. The return travel shows the wire traveling over and under small-diameter rolls, which keep the wire under tension and help guide it. The wire guide roll is installed such that one end is stationary and the other is free to move back and forth in the horizontal plane. The horizontal movement of one end causes the roll to be skewed, or set at an angle to the direction of travel of the wire. This angle creates a sideways force on the wire and can move the wire toward one side of the machine or the other. As the wire runs on the machine, it may stretch or wear unevenly and start to move toward either the front or back of the machine by itself. The wire guide roll can be installed with a simple wire position sensor to sense and automatically correct the movement of the wire to keep it properly positioned on the machine.

Figure 6.11 also shows whitewater trays below the top run of the wire. The water removed will contain fibers and other materials and needs to be collected for reuse. Furthermore the whitewater must be kept from the returning wire. It is necessary to clean the fibers and other materials from the wire on the return run. Although not shown in the drawing, there will be at least two water collection pits under the wire. The largest pit is found under the major portion of the wire and collects the wire showerwater. This water is usually quite contaminated and cannot be reclaimed. The other pit is just under the couch roll and is called the *couch pit*. The couch pit is where the web is dropped during startups or when the web breaks. During normal operation, the edges of the web are also trimmed off with water jets and these edges are constantly dropped into the couch pit. The water from the couch pit will therefore contain fibers and will be returned to the stock prep area for reuse.

The drawing in Figure 6.11 shows the use of a *dandy roll*, an open roll covered with a wire fabric similar to the forming wire. The dandy roll is used to press down loose fibers, to make the top surface a little flatter and possibly to put a watermark on the paper. The watermark gives the paper a distinctive or permanent identification and is used in the manufacture of money or bond papers to prevent counterfeiting, or just for advertising purposes. The dandy roll is an optional device and is not used on some grades of paper.

Water Removal Devices and Operations

The equipment under the wire is designed to aid in the removal of water and to support the wire. The first piece of equipment shown in Figure 6.11 is the *forming board*. The forming board can be simply a series of boards with slots between them to allow the water to run down into the whitewater tray. At low consistencies, just out of the headbox, water drains from the web freely. As the stock travels down the wire and water is removed, the consistency increases and the removal of water becomes more difficult.

The majority of water is removed in this section with the aid of *table rolls* or *hydrafoils*. The two are compared in Figure 6.12. Table rolls are small-diameter rolls (3-10 in. dia.) supported on each end with bearings, which allow them to rotate. The rolls are driven by the wire passing over them. As shown in Figure 6.12, the wire carries the stock over the roll and the combined motion of the wire and the roll creates a vacuum on the back side of the roll. The diameter of the roll and the speed of the wire are both important factors in determining the amount of vacuum. As the speed of the machine increases, the water will be carried farther around the roll until it is thrown back up against the underside of

Figure 6.12. Comparison of table rolls and foils

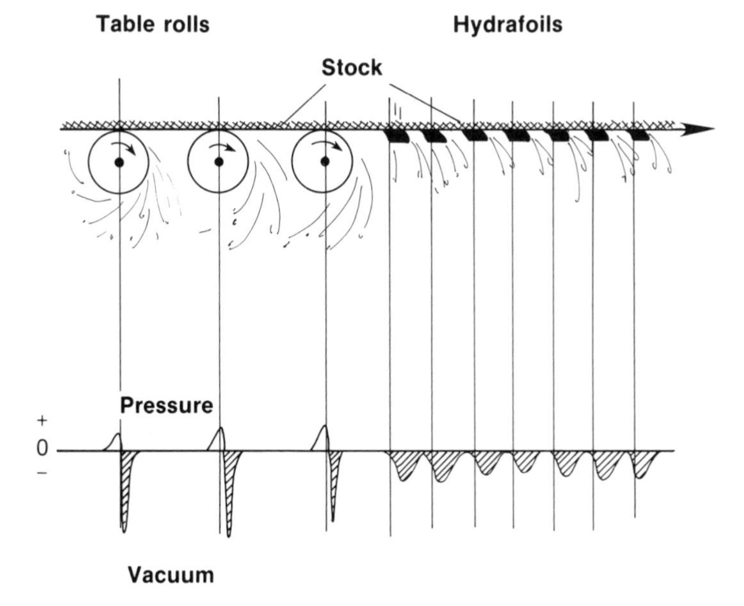

Cylinder Former Designs

An early cylinder former is shown in Figure 6.14. It takes its name from the fact that the wire screen is wrapped around the surface of a cylinder. The wire-covered cylinder is rotated through a vat containing the diluted fibers and the web is formed on the outer surface of the cylinder as the water passes through the wire and flows out the end of the cylinder. By rotating the cylinder in the vat, a continuous web can be formed. The web is removed from the wire by a porous woven blanket called a *felt*. As the wet felt is pressed against the wet web, the web will adhere to the less porous of the two surfaces. Since the felt is less porous than the wire, the web is removed from the cylinder. The solid roll behind the felt is called the *couch roll*. The name comes from the ancient hand-made operation in which the sheet was couched off the wire onto the felt.

The early cylinder formers came in two basic designs: the *direct flow vat* and the *indirect* or *counterflow vat*. In the direct flow or "Uniflow" vat, the flow of the stock moves in the same direction as the cylinder; in the counterflow design, it moves in the opposite direction. Both types are plagued by problems with flow control and poor formation. Formation of the web begins as soon as the wire enters the vat. As the fibers build up on the wire, the flow rate through it and the web decreases. The later layers of fibers are therefore not deposited randomly, but rather are influenced by the condition of the web already formed. This effect, referred to as *secondary* or *tertiary formation*, is believed by some to be the cause of the poor formation. Another theory suggests that clumps of fibers are actually lifted from the surface of the web and redeposited as clumps rather than as individual fibers, causing lumps in the web.

One modification designed to improve formation is the *dry vat* conversion shown in Figure 6.15. The counterflow cylinder is easily modified by the positioning of a dam (simply a piece of rubber nailed to a board) at about the 5 o'clock position in the forming vat. The stock then enters the vat and begins to flow through the wire, forming the web. The whitewater that passes through the wire flows back through the wire and out the bottom of the vat, as shown in Figure 6.15, or else is removed from the end of the vat as is normally done with counterflow cylinders. By eliminating the pass through the vat of stock, the lumpy for-

Figure 6.14. Uniflow cylinder former

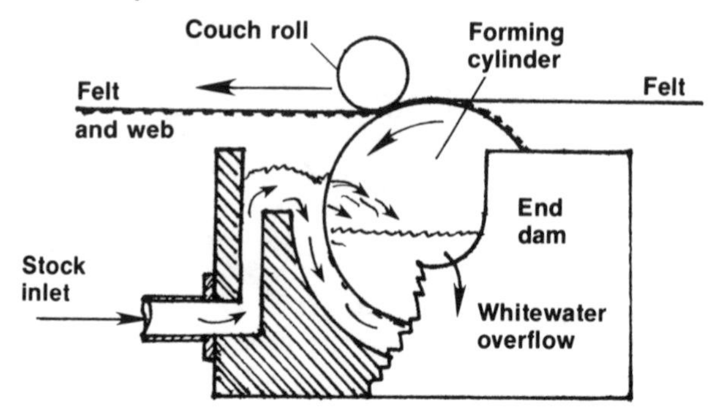

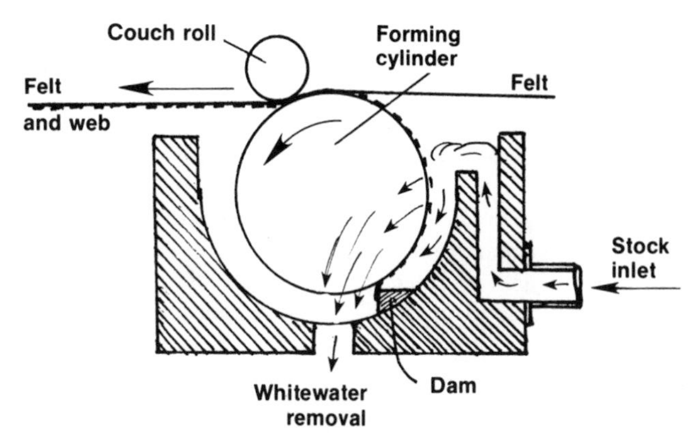

Figure 6.15. Dry vat conversion of a counterflow cylinder former

mation is avoided. However, the formation may still be wild if there is too much turbulence in the forming zone.

Another form of cylinder former modification has been through the use of headboxes with the cylinder. The shape of the headbox is quite variable and not the same as described for fourdrinier machines. Figure 6.16 shows one type of former using a flow spreader-type approach flow and headbox that conform to the cylinder surface. To aid drainage, suction is frequently needed inside the cylinder. The former shown in Figure 6.16 has a rotating outer shell and stationary suction quadrants inside. Figure 6.17 shows another approach to the suction problem: sealing off both ends of the cylinder, drawing a vacuum inside and removing the whitewater from the end of the cylinder. The former in Figure 6.17 also shows a different style headbox more similar to the ones used with fourdrinier formers.

Figure 6.16. Suction former

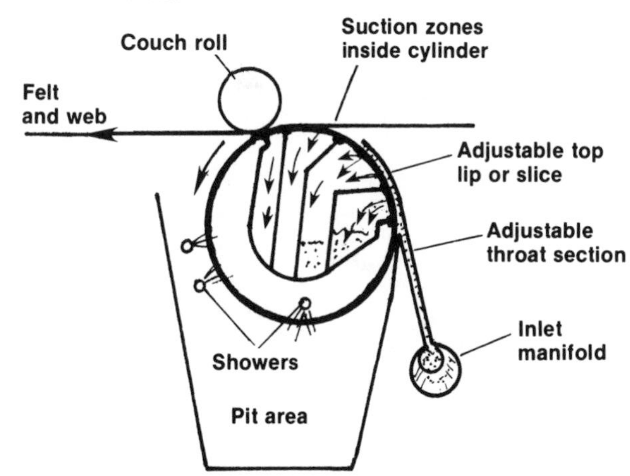

A third approach is through the use of a pressure former. The former shown in Figure 6.18 delivers the stock to the wire surface while it is still inside the top lip of the headbox. The stock in the headbox is pressurized to help force the whitewater through the wire, reducing the need for suction. However, the cylinder may still be evacuated to help speed formation or increase the flow rate through the wire.

The incorporation of these forming devices into cylinder machines was discussed earlier in this chapter in the overview and will be discussed again in Chapter 10. Some of the cylinder former applications are being replaced by twin wire formers, which are discussed in the following section and also in Chapter 10.

Twin-Wire Formers

Although some new single-wire fourdriniers are still being made, and a large number of older machines are still making excellent quality paper, there has been a steady trend toward the use of twin-wire machines in all paper grades. It was reported in 1980 (Thorp, et al., TAPPI) that the trend was toward twin-wire machines, although the speed and quality advantages had not been clearly established. Developments since then seem to indicate that there are advantages for the twin-wire, since the new machines are primarily of this type. The development of better headboxes, foils and synthetic forming fabrics continues to add to the advantage for the twin-wire machine. There are three major categories of twin-wire machines based on the method of water removal and the overall machine design; (1) roll formers, (2) blade formers and (3) hybrids or combinations of these two and/or flat fourdriniers.

Figure 6.17. Conventional Ultra Former design

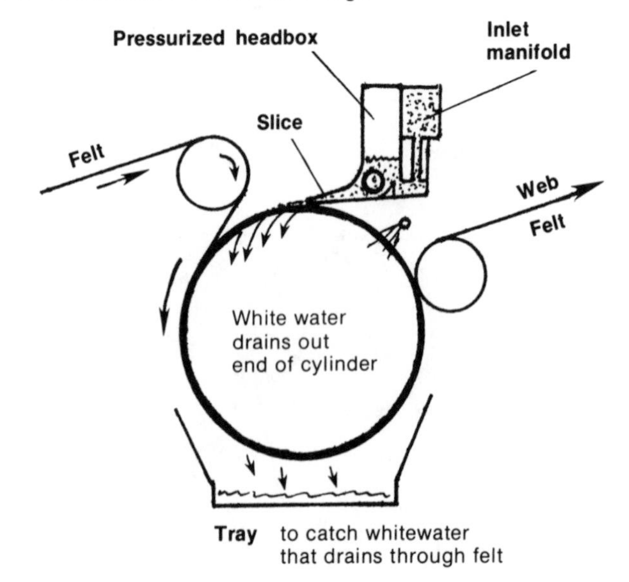

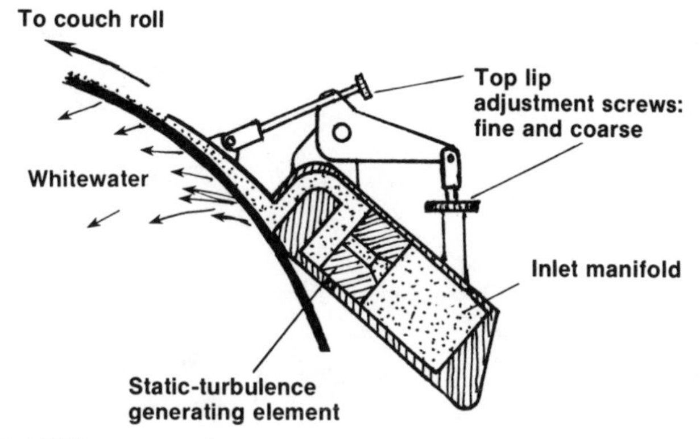

Figure 6.18. BRDA pressure former

Figure 6.19 shows the use of rolls around which the wires are led, so it is clearly of the first category. Some which only wrap around one roll, will be shown in the chapter on tissue formers. The water is removed by a combination of centrifugal force and the pressure developed by the tension on the outside wire. These formers are claimed to have lower power requirements, longer fabric life and high-speed capabilities, but may not give the best formation and internal bond. An example of the blade type former is shown in Figure 6.20, where a series of foils is used to remove the water. The pressure pulses generated by the

Figure 6.19. Papriformer design

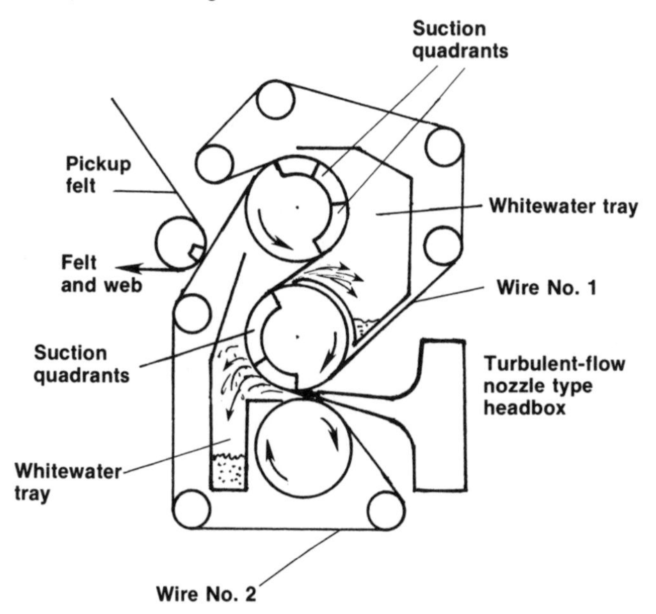

A Beloit twin-wire machine of the type shown in Figure 6.21.

foils can cause more turbulence in the layer of the stock next to the wire, causing a less gentle water removal and possibly less retention of fines and fillers in that layer. The higher power requirement is due to the dragging of the fabric across the static elements. The better formation reported for these formers could be due to lower consistencies in the headbox, or the longer time for water removal. Depending on the radius of the forming roll and placement of the blades, the roll former should be able to remove the water twice as fast as a blade former and ten times as fast as a flat fourdrinier. However, formation control is a rather complex problem and cannot be simplified that easily. The blade former shown in Figure 6.20 also obtains some water removal from the fact that the two wires are led through a large-radius arc.

The combination of blade and roll former is shown in Figure 6.21, where the wires are wrapped around a roll for less than 90 degrees, then led through some

Figure 6.20. Bel-Baie design

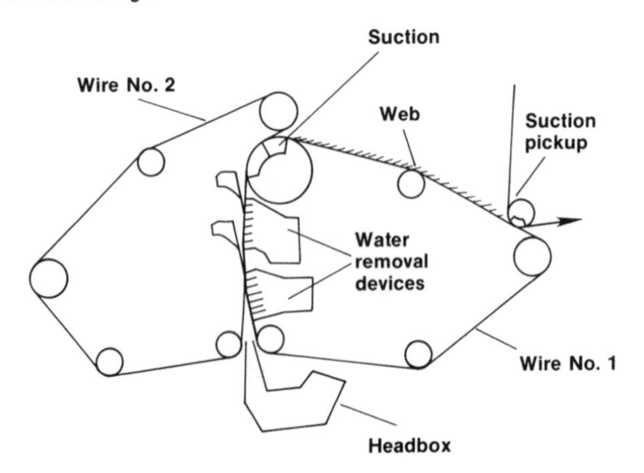

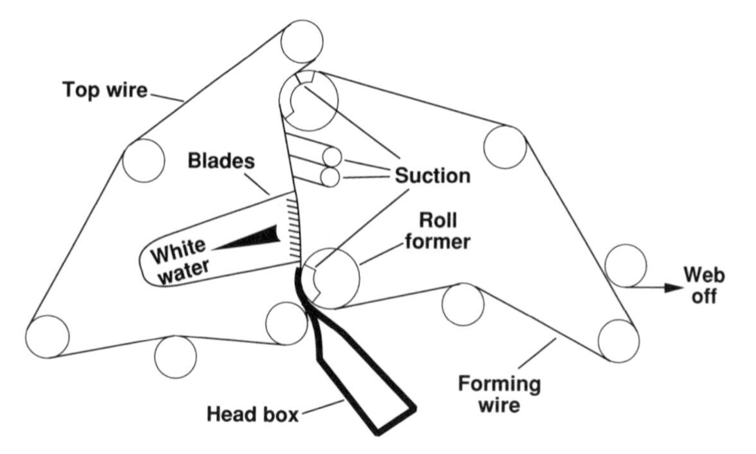

Figure 6.21. Roll and blade-type twin-wire former

blades for further water removal. It should be noted that the blade former in Figure 6.20 also shows the wires wrapping a roll after the blades, but this wrap is not generally considered to be a major portion of the water removal. The wires need to be wrapped to redirect the web and aid in removing the web from the wire. Any twin-wire machine needs to be designed to cause the web to follow the desired wire and not cause undue stress on the wet web as the wires separate. The web may be caused to follow one of the wires by using a suction box in the roll or in a shoe ahead of the roll. Other combination machines have been created by rebuilding flat-wire fourdriniers through the addition of a second wire. The Bel-Bond, show in Figure 6.22, is one of the more common of this type.

At this time, the roll formers are more likely to be found in tissue applications and the blade formers in printing grades. But, with the exception of the seeming

Figure 6.22. Bel-Bond secondary headbox design

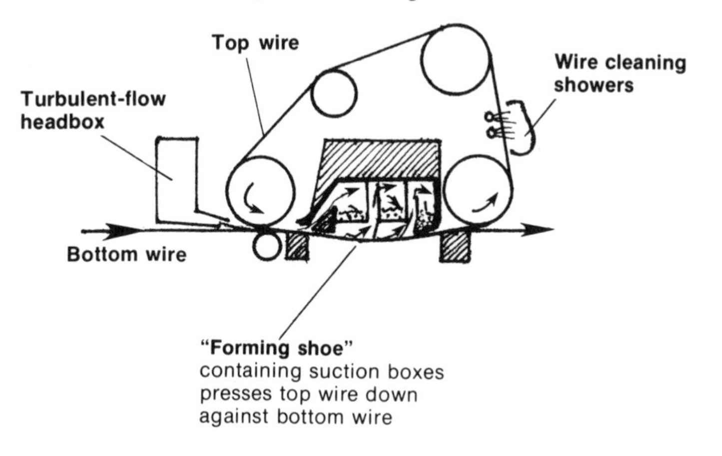

strong even though bonding has started to take place. The press section will be operated to remove as much water as possible to reduce the load on the dryers that follow it, and will also be adjusted to control the strength of the web, if possible. The main consideration in the operation of the press, however, is water removal.

DRYING THE WEB

The web of paper can only be raised to about 36% to 40% consistency by pressing; the rest of the water in the sheet must be removed by evaporation. The conventional method for evaporating the water has been to pass the wet web around steam-filled cylinders, which heat the web to the vaporization temperature of the water so that the water will evaporate. The cylinders are called *dryer drums or cans*. They are normally about 4 or 5 ft in diameter and as wide as the machine.

The dryer cans are mounted in two horizontal rows such that the web can be wrapped around one in the top row and then around one in the bottom row. The web travels back and forth between the two rows of dryers until it is dry. There are two notable exceptions to the use of two rows as described. One exception is the use of one roll about 12 to 15 ft in diameter, which dries the web in one pass. This single-roll dryer is called a *yankee dryer*, and is used to dry lightweight papers such as tissue. A second modification of the two rows of dryers is the use of *vertical stacking* in combination boxboard or cylinder machines, as shown in Figure 6.3. In the boxboard machine, the number of dryers needed is so great that the rolls are frequently stacked one above another, with the paperboard web weaving back and forth between the two stacks until it reaches a height of about 10 dryer cans. When the web reaches the top of one pair of stacks, it is passed to another pair of stacks, where it weaves back and forth around them until it reaches the bottom. The web will go up and down perhaps a dozen stacks of dryers before it is dry. The number of dryer cans needed in any machine is determined by the basis weight of the paper, its moisture content coming into the dryer section and the speed at which the machine is to be run.

Basic Drying Theory

The basic theory of drying can be discussed in terms of a combination of drying rate, temperature and moisture content, as shown in Figure 6.26. The initial phase of any drying operation is to raise the material to be dried to the evaporation temperature. This first stage is called the *warm-up zone* or *increasing-rate zone*. We can see in Figure 6.26 that as the temperature is increased, the rate of evaporation increases, but the moisture content does not change too much during this phase. The first three to six dryers in any paper machine are primarily devoted to raising the temperature of the sheet. The web coming from the press section will usually not be much above room temperature, if it is even that warm. The paper machine may use hot water in the forming section to increase drainage rates

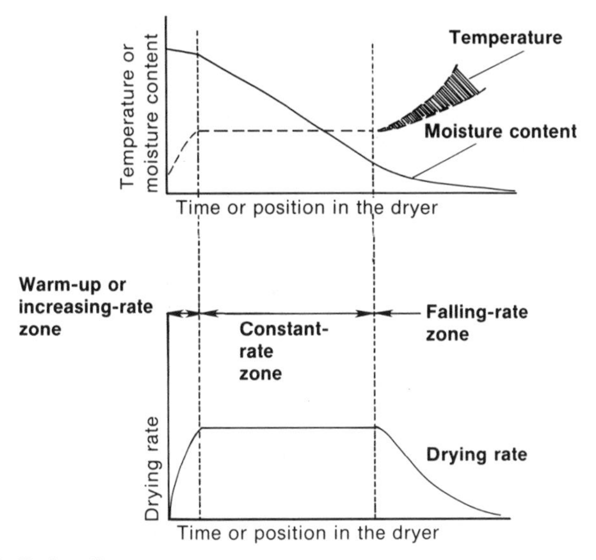

Figure 6.26. Drying theory

there and to aid in the removal of water in the press, but the evaporation of water from the web as it passes through the presses and into the dryer will cool the web rapidly, making it difficult to maintain the web at elevated temperatures. In order to warm the web yet prevent localized overheating of the web, the temperatures of the first dryers are usually around 65°C (150°F). If the web is placed in contact with a dryer can that is too hot, the fibers on the surface of the web will dry and stick to the dryer surface. The fibers then are pulled from the web, disrupting the web surface and causing a linting problem on the first dryers. The temperatures of the dryers are controlled and raised to more than 100°C (212°F) as soon as possible. Once the web has been warmed, the dryer temperatures may be allowed to go as high as 170° to 200°C (350° to 400°F). The temperature of the web, however, will not rise above the evaporation temperature of the water as long as there is water in the web.

The evaporation of water from the web cools the web and prevents it from becoming as hot as the dryer cans. This constant-temperature zone is also called the *constant-rate zone.* Once the material to be dried has been raised to the maximum temperature, the limiting factors in the rate of evaporization are: the surface area of the water that can be exposed to the atmosphere, the speed with which the water vapor can be removed from the water-air interface and the rate of heat transfer into the web. Since the surface area is determined by the size of the machine, there is little to be done about that parameter. Improving removal of the water vapor will be discussed later. The rate of heat transfer is dependent on the temperature differential, among other things, so the dryer temperatures are kept as high as possible. Once the water level in the web falls below the level that allows easy migration of the water to the surface of the web, the drying rate is reduced. It was assumed in the earlier discussion of drying rates that the water web is allowed to dry in contact with the can, it takes on a flat, smooth surface like that of the dryer can. This application is used for grades such as carbon papers, which need to be very thin and have a flat surface. These are called machine-glazed grades.

Another grade made on the yankee is creped tissue. By scraping the web from the dryer can just before it is completely dry, the web becomes wrinkled in the cross-machine direction, a condition known as *creping*. Tissue is not the only grade made in this manner; decorative crepe paper, crepe wadding or packing paper and a few other grades are also made with the yankee dryer. The creping of the web causes the web to become shorter in the machine direction, so subsequent sections of the machine need to be run slower to prevent pulling all of the crepe out of the paper. The crepe gives the paper added softness, bulk and absorbency, which is why it is popular for tissue. However, not all tissue is creped. Some is made on a yankee without creping, or may even be made on a conventional dryer and then be made softer by embossing. Embossing is a form of surface modification that will be discussed in Chapter 7.

Since the paper must be dried in one pass around the yankee dryer, it is necessary to supply additional energy above that provided by the dryer can. The additional heat is supplied by hot air blown on the surface of the web from hot air caps around the yankee dryer. The hot air is blown at relatively high velocities directly at the surface of the web. The high velocity breaks through the layer of air and steam at the surface of the web to improve the heat transfer and help remove the steam from the web. Yankee dryers will be discussed in greater detail in Chapter 11.

COMPUTERIZATION AND PROCESS CONTROL

The continuous nature of the operations performed in the paper industry has allowed the use of process control devices and ultimately computerized control. At this time the major operations in both pulping and papermaking are under continuous computer control, and many of the converting operations are being added to the list. We will first discuss some of the basic elements and techniques used for process control, then indicate the applications currently employed in the different operations.

Fundamentals of Process Control

In order to control a process, it is necessary to measure input and output properties of the process. Devices for measuring the properties are commonly called *sensors*. We also need to have input parameters for the process that can have an effect on the measured output property, and we need to be able to control them in some manner. The actual devices used to control the input parameters vary widely, and may be called actuators, controllers or transducers.

Transducers is actually a rather broad term that is used to indicate that a signal of some sort is being transmitted and usually converted to another form of signal. The term *controller* has been used to refer to the total control system, including sensor, actuator and controlling logic. A simple control system would be a person turning on the water, holding his hand in the stream flowing from the faucet (sensor), deciding if the temperature is right (control logic) and adjusting the valves to achieve the desired temperature. The temperature of water delivered from a water heater is controlled similarly: a sensor is installed in the exit pipe to measure the temperature; that value is compared to a set point in some form of mechanical or electrical device to determine if there is a difference; and a signal is fed to a heater or steam valve to adjust the temperature.

This simple system becomes more sophisticated and complex depending on the type of control action. If we have only the ability to turn the heater on or off, we will probably see the temperature fluctuate above and below the set point based on the process lag or time delay required for the corrective action to find its way to the sensor. The fluctuations can be dampened by using a proportional control, which allows the amount of corrective action to be adjusted by the distance of the output signal from the set point. This type of control requires an adjustable input control actuator and a controller with that type of logic capability. All of the operations described to this point have been available for years in the form of pneumatic controllers. These require that the sensor incorporate a transducer to convert the signal to a variable air pressure; the pneumatic controller itself, which has provision for mechanical adjustments to set the control logic; and finally, another transducer to convert the air pressure output to an action that can control the input parameter.

All of the control discussed to this point falls in the category of *feedback control*: the property is measured after it happens, and corrective action is taken on the material starting into the process after the measurement was taken. This type of control is not bad for systems that are slow to change and have fairly short process delays. In the case of a paper machine operating at close to a mile a minute, where the paper goes from the wet end to the reel in a matter of seconds, there are some feedback controls that are useful. This type of control is not satisfactory for a continuous digester: the chips may take an hour to pass through, and any corrective action will require another hour to show up and assumes that the chips being fed now are the same as those fed an hour or more ago. What is needed is *feed forward control*. This type of control requires that the properties of the input feed(s) be measured and a predictive equation exists, which can calculate what the probable resulting output parameter(s) will be in order to adjust them in advance. With the development of computers to speed the calculations needed for feed forward control, this type of control has become prevalent. The newer systems incorporate a combination of feed forward control and feedback monitoring of the accuracy of the predictor equations, with the ability to adjust the predictive equations to better control the system.

The Development of Computer Control

Early mill control consisted of a large number of controllers operating on individual components of the total process. Many of these controllers included recorders to give a permanent record of the processes. These were collected and used by the management to monitor and record mill operations. Early computer installations were able to replace some of the control (feedback) loops and streamline the record gathering and data distribution. These computers were large mainframe computers typical of the late 1950s and 1960s. As newer and better sensors were developed, more operations could be monitored and controlled, and the capacity of these early computers was soon exceeded.

Due to the hostile environment of the mill and delicate nature of this generation of computers, the computer had to be housed in a controlled atmosphere with miles of shielded cable feeding data and control signals to and from the computer. The advent of the smaller microcomputer in the 1970s allowed the use of several computers at different locations in the mill, usually connected to a main computer for management data logging, inventory control and accounting functions. Continued improvements in the development of sensors, control strategies and equations to predict the performance of the processes and process machinery has continued to increase the demand for computer control. At the same time the computer has become smaller yet more powerful, more tolerant of the hostile environment of the mill and more dependent on software than hardware.

At the present time we find at least two different mill control strategies: (1) Individual processes controlled by individual minicomputers, located at or near the process being controlled, which are linked by network systems that may use telephone lines rather than the dedicated hard-wire-shielded cables of the early days. These individual controllers report to a ma ster computer, whose main function is as databank. It is usually controlled at the management level and interfaces with the accounting functions. (2) Distributive control also has a large number of linked individual minicomputers located at or near the individual processes. But it has software that enables the same control logic to be used by the individual processes (with the necessary modifications suitable for each process), and for any of the processes to be accessed from any of the computers. This is a more highly interactive and flexible system; in fact, is felt by some to be too flexible and complex.

Another feature of the modern computer control system is the use of computer data logging in the testing lab. Many of the routine test instruments have been adapted to transducers, allowing a signal to be fed directly into a computer for data logging, report generation or even use in the development of better control logic. At the same time, the test instruments have been modified to allow automatic feeding of samples through the instrument and, in many cases, into adjacent instruments. It is possible for the machine operators to remove a sample from the reel at the turnup, take it to the lab and feed it into the first of a row of connected test instruments. These instruments will measure the optical properties, weight, moisture, thickness, smoothness, porosity and strength of the paper in a matter of minutes. This data can be made available at the same time (depending on the network system used) to the operating personnel, management or anyone connected to the system.

Although most of the discussion to this point has been directed toward the control of a continuous operation, the computer is perhaps even better at controlling operations as they change. Man of the early computer applications were justified on their ability to control the many complex operations needed to change basis weight or grades on a paper machine. Under manual control, the operating staff needs to physically adjust valves and other controlling parameters in the proper order and to the correct amount to affect the desired change. These changes are made in a certain sequence based on the operators' past experience, and are monitored visually by the staff during the changeover to fine tune or adjust the changes in controls, attempting to hit the desired new grade specifications as quickly as possible. The computer can utilize past changeover data to predict the new set points, be programmed for the necessary time delays or be instructed to monitor certain properties for the appropriate time to make additional adjustments. Sensors can be located at several parts of the process so that the operating staff does not have to run all over collecting and coordinating data.

This type of capability is as useful in batch pulping operations or the preparation of coating batches as it is in monitoring and changing the continuous operations. The accounting functions mentioned earlier include the ability to know when enough paper has been produced to fill the order, which orders can be run together, or which sequence will run best, and monitor the cost of operations continuously.

The operation of a modern mill without computer control is unthinkable. Even the design of a new mill can be performed with a computer-aided design (CAD) system. The entire mill system can be designed and built using computer-generated plans and three-dimensional views to check the ease with which the system can be operated as well as to train the operators.

Systems Being Controlled

Pulping and Bleaching

Inventory control begins with manual entry of data on wood received, and possible incorporation of automated weight data entry, and continues through the entire mill. Chips can be measured for moisture and density, as well as flow rates and volume control in storage vessels. It is therefore possible to monitor, control and input data on the feed flows to digesters to modify and control cooking times and temperatures to suit the chip feed. During the cooking and subsequent bleaching operations, samples of the liquor being recirculated through heaters can be monitored to further modify the cooking conditions. The liquors can be measured for temperature, pH, density and some chemical compositions. Feeds to recovery boilers can be measured and monitored, as can other flows in the recovery systems. Control strategies exist for feed forward control of most of these systems, either as continuous or batch operations.

When the fibers have been converted to a slurry, the consistency can be monitored to further contribute to inventory control as well as calculate chemical needs in bleaching operations. This data is also essential to the control of the refining operations. The feed forward control of refining is frustrated by difficulties in the complete understanding and control of the many parameters contributing to the refining operation. Furthermore, the results of refining take different forms (cutting, bruising and so on) and make the actual measurement of the degree or type of refining action difficult. On-line freeness measuring devices have been available for some time, but freeness is not an exact measurement. Methods for measuring fiber length have been developed and are now being adapted to at least periodic measurement if not continuous monitoring. The earliest control of refining relied on adjusting the flow, consistency and plug setting in response to temperature rise, freeness, drainage on the wire and power consumption, or combinations of the above. Some of these same control strategies are still in use, but are now interfaced with the computer control.

Papermaking Operations

Flow rates and consistencies of most flows going toward the headbox are measurable, allowing feed forward prediction of jet velocities, basis weights and other properties. Actual properties of the paper that can be measured on a continuous basis include basis weight, moisture, caliper, porosity, smoothness, brightness, opacity and color. The strength of the paper is more difficult to measure, since most strength measurements are destructive in nature. It is possible to monitor sonic transmission through the web, which is related to the modulus and therefore strength, but this is not a universally accepted practice. Other measurements are made in specific grades or mills, but the above listing covers the most widely accepted properties. Each can be tied to a control strategy to allow closed loop control, either in a feedback or feed forward mode, with updating or modification of the control strategy based on the actual measured properties.

A sampling of some of the control loops includes control of the rush/drag ratio or relative velocities of the stock coming from the slice and the wire; filler content to control opacity; chemical feed rates to control retention, color and other optical properties; steam feeds to control moisture content; and speeds and flow rates to control basis weight. Many of the properties can be measured with a traversing head to allow the development of a profile or cross machine plot to show the uniformity of these properties.

Not all can be controlled with that type of accuracy. Moisture profile can be controlled either through using variable infrared heating across the web, or through control of water temperature across the web with steam boxes on the wire or in the press section. The caliper can be adjusted by varying the temperature of the rolls in the calender stack to vary the degree of compression of the web as it passes through the nips. The cross machine control of basis weight and caliper through adjustment of the slice has been accomplished by hand for years, but the development of servo motors or heated rods to make these fine adjustments has only recently become widespread. The slice adjustment problem is complicated by the complex nature of the hydrodynamics in the headbox. When the flow is restricted in one region it changes the flow in the neighboring regions. Since the slice is a continuous thin strip stretching across over 200 inches of the machine, an adjustment at one point causes changes in stresses at other points. This can cause additional adjustments to be needed. Furthermore, the change of the slice opening will not only change the thickness of stock flowing through, but may also change the velocity and have harmful effects on the rush/drag or other properties.

Converting Processes

Computer controls exist for most of the web conversion processes for tension control and other monitoring functions. For pigmented coating, the coating mixture can be batch or continuously prepared by various computer control schemes. Meters must be specifically designed to measure some of the coating ingredients, or ingredients can be measured together by weighing the mixing tank. Viscosity and density or solids level of the mixture can be measured and solids level adjusted as a control parameter. The coat weight on paper can be measured using beta ray transmission, the same sensor technology used to measure the basis weight of paper. Boxboard coat weight is more difficult to measure because of the greater weight of the uncoated substrate and resulting small difference after coating. Moisture content of the coated material can also be measured and controlled. The individual applications for computer control in converting operations are numerous, but too specific to detail in this introductory treatment.

REFERENCES

Justus, E. J., *Pulp & Paper* 52(12):183-184 (October 1978).

Kallmes, O. J., *Tappi* 62(3):51-54 (1979).

Kocurek, M. J. Ed, *Pulp & Paper Manufacture*, Third Ed. Vol. 7 The Joint Textbook Committee of the Paper Industry, Atlanta, TAPPI (in press).

Lavigne, J. R., *Instrumentation Applications for the Pulp and Paper Industry*, San Francisco, Miller Freeman, 1979.

Pulp & Paper, "Multi-ply formers," 52(14):89 (December 1978).

Schmidt, S., *Paper Age*, January 1980.

Smook, G. A., *Handbook for Pulp & Paper Technologists*, The Joint Textbook Committee of the Paper Industry (TAPPI, Atlanta, and CPPA, Montreal).

7 Web Modification

The operations included in this chapter are more commonly considered to be part of the papermaking operations, but have been placed in a separate chapter to reduce the length and complexity of the preceding chapter. The papermaking operations could proceed quite well without these modifications, but the quality or usefulness of the paper or paperboard would suffer. The operations have been categorized by type and location rather than by their effect on the properties of the final sheet.

INTERNAL METHODS OF MODIFICATION

Sizing: Improving Water Resistance

Since paper is made with water as the carrier and as an aid to bonding, paper is extremely sensitive to the reintroduction of water into the fiber network. Cellulose is very hygroscopic and will absorb water from the air. The water will cause the sheet to swell and perhaps curl, as discussed in Chapter 2. Therefore it is important to treat the paper to improve its resistance to water. There are two major methods of obtaining water resistance: the addition of chemicals during web formation, called *internal sizing,* or surface application of chemicals after web formation, called *surface sizing.* The internal method will be considered here; surface sizing will be discussed later in this chapter.

The most common, and oldest, method of internal sizing is with rosin and alum. Rosin, a natural organic acid obtained from trees, is emulsified in water and added to the fibers before they are sent to the paper machine. The rosin is slightly anionic and will tend to stick to the fibers. After the rosin has been mixed with the fibers, alum is added to the stock. Alum is a water solution of aluminum sulfate, with some of the aluminum also in the form of aluminum hydroxide. The alum also has enough extra acid to be at a low pH and will be added in sufficient quantity to lower the pH of the stock to about 4 to 5. The alum flocculates with the rosin and with itself, creating flocs that adhere to the fibers. The rosin-alum flocs are water resistant after drying, and their presence helps the web resist water penetration. The degree of water resistance obtainable is variable based on the level of addition, pH and some other variables, but in any case is not enough to make the paper truly waterproof. Water will still penetrate into the web, but at a slower rate. Furthermore, the web can still be broken up by the combined action of water and mixing energy.

The alum remaining in the web is harmful to the permanence of the paper. If the paper is used in documents that must be kept for a long time, the rosin-alum sizing system can cause the paper to become brittle. Early handmade paper that has lasted so long was probably sized with animal glue. Recent developments by chemical companies have given the papermaker several alternatives to acid sizing systems. There is currently pressure on the paper industry to produce alkaline and neutral sized papers to increase their permanence, and the chemical suppliers have responded to this demand. We now can select from rosin-based systems that can perform at neutral or nearly neutral pH or other organic compounds that can be used without the addition of alum or any other acid.

Wet Strength and Bonding Additives

Sizing materials can give the paper only some small degree of water repellency. If it is necessary for the paper or paperboard to be wetted and still retain strength, a different type of additive called a *wet strength agent* must be used. The wet strength agent is added to the stock prior to formation of the web, the same as the sizing agent. By adding the wet strength agent to the stock prior to web formation, it can adhere to the fibers and also be deposited on the bond areas during web formation. The wet strength agent then functions in the web to protect the bonding and also to help hold the fibers together when the web is wetted. Wet strength agents are important in: paper toweling, wrapping or bag papers that may be wetted during use and paperboard that will be used in wet or damp applications.

Starch is also added to the stock before web formation to help glue or bond the fibers together. Starch is chemically quite similar to cellulose and can bond to fibers to increase the degree of bonding in the web. The presence of the starch in the web may improve the water resistance a little; however, starch is less water resistant than cellulose and therefore will not appreciably improve wet strength. The penetration of water into the web is dependent on the pore structure of the web as well as on the contact angle of the liquid and the fiber surface. If the starch fills some of the voids in the web, it can slow the penetration of water either into or through the web and thereby help with water repellency. Other vegetable and animal gums, which may also contain proteins, have been used to help develop dry bonding, but starch remains the most common additive.

Optical Property Modification

Colored paper has traditionally been made by the addition of dyes or colored pigments to the fibers either prior to or during the formation operations. There are three classes of dyes commonly used with paper: acid, direct and basic. These dyes differ in their properties of tinctorial strength, permanence, affinity for fibers and cost. *Basic dyes* are quite strong and have affinity for unbleached fibers, which makes them best in that application. *Acid dyes* work better with bleached stock, but are of lower strength. *Direct dyes* attach to the fibers and therefore seem to have a higher tinctorial strength. Any dye will be partially lost to the whitewater during web formation. The concentration of dye in the whitewater varies with the amount used as well as the type. Any dye in the whitewater is a problem to the mill, since the dye can find its way into the mill effluent stream. Even if the whitewater is retained in the mill to be reused, the colored whitewater can be used only in that color paper and must be flushed from the system when changes to other colors are made. Making colored paper by the internal addition of dyes is extremely troublesome to the mill and requires that the cost of such products be high enough to compensate for those difficulties. Many mills have developed methods for surface application of dyes to solve the pollution problems.

Mineral pigments, both white and colored, can also be added to the fibers prior to web formation to modify the optical properties of the web. Colored pigments can obviously be used to color the paper or paperboard, but can also contribute to problems with contamination of the whitewater. White pigments are the most commonly used. White pigments will improve the brightness and opacity of the paper, but may reduce the strength since they replace some fibers in the web and will also interfere with bonding between the fibers. Cellulose fibers are essentially transparent to light and will form a clear film of cellophane when dissolved and redeposited in a film. Paper is visible and obtains its optical properties from the many light-scattering surfaces in the web structure. Glass or ice is also transparent in single pieces, but when it is crushed, many surfaces are created that will scatter light, making the glass or ice visible and perhaps even opaque. White pigment fillers may also be transparent crystals but since they are composed of fine particles smaller than the fibers, an equal weight of pigment will give a substantial increase in scattering within the web.

The most common pigment fillers are clay, calcium carbonate and titanium dioxide. The *clay* used is the kaolin crystalline form, which is mined in many places throughout the world, comes in a variety of particle sizes and is by far the most commonly used pigment. Clays are fairly bright, although not much brighter than some fibers, but are the lowest cost filler. *Calcium carbonates* are quite a bit brighter, but cannot be used with acid sizing systems since the acid will decompose the carbonate, releasing carbon dioxide and causing foam. The carbonates still find use with nonacid sizing systems and in cigarette papers, where they help control ash development. *Titanium dioxide* is the least used, partially because it costs about 10 times as much as clay, and also because it is so effective that little is needed. Titanium dioxide has a distinctively different index of refraction and therefore is most effective at developing opacity in paper.

The level of substitution of filler for fiber can run from as low as 2% for titanium to as high as 30% for clay or carbonate. Beside their effect on brightness and opacity, the fillers will also change the printing and handling properties of the web. The filled sheet will usually print better due to better formation, smoothness and control of ink penetration into the web. Individual pigments or mixtures of pigments are used to maximize final sheet properties. There is, however, a limit to the amount of filler that can be used without harmful loss of strength, stiffness or other performance properties of the web.

SURFACE MODIFICATION METHODS

Chemical Modification: Surface Sizing

The web can be treated with starch or other gums at the size press to help improve water repellency. The mechanism of action is either to change the contact angle of the water with the paper surface or to plug or seal pores in the web surface. Starch is the most common surface sizing material. Since starch is not too water resistant itself, it is believed that it functions primarily by improving bonding at the surface, to plug the surface and retard penetration of water. Polyvinyl alcohol, guar gum and animal gums are also used, and are claimed by some to be more effective.

Surface sizing can also create a surface that will resist erasures or other forms of abrasion. Although the main mechanism is one of improved bonding, fluorocarbons, silicones, waxes or other "release" agents are also helpful in preventing the ink from penetrating too far into the web, making it easier to remove. Special chemicals are sometimes applied in an attempt to develop resistance to oils, grease or other materials, but it is not easy to develop a high degree of resistance with a simple size press application. A high degree of resistance requires heavy application of materials in a conversion operation, and goes beyond the scope of this chapter.

The size press is found in several designs, as indicated in Figure 7.1. The vertical size press (A) is the oldest and is capable of applying the chemical solutions discussed to this point. The solution is applied to the top roll. The excess will flow past the web into the pan, where it will be picked up and applied to the bottom of the web by the bottom roll. The other modifications were developed to allow the inclusion of pigmentation into the size solution or to allow coating of the paper. A range of surface applications are being applied today from the pure solutions discussed so far through pigmented coatings, which will be discussed later. As the amount of pigmentation in the solution increases, several problems develop. The specific gravity of the coating caused the web to sag in the vertical

Figure 7.1. Size press coaters—A: Vertical; B: Horizontal; C: Inclined; D: Gate roll

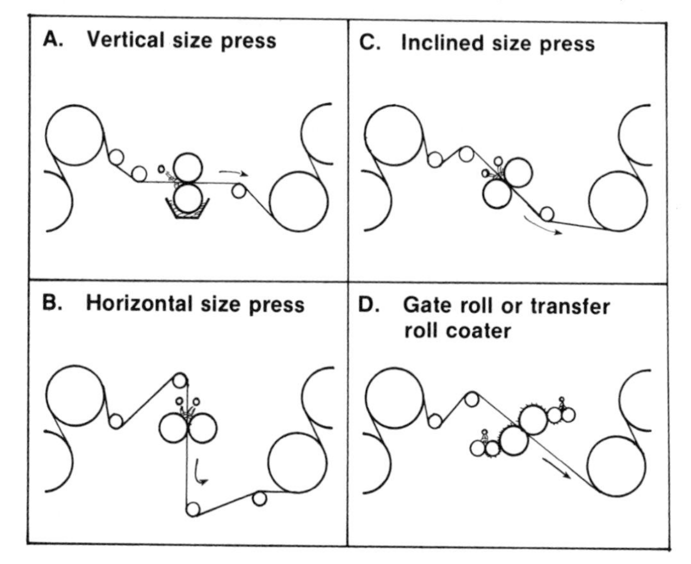

size press. The horizontal press must deal with the wet web heading to the floor at a high rate of speed. The inclined size press is a reasonable compromise for the problems described so far. The horizontal and inclined also have problems with coating splashing in the pond at higher speeds. All designs experience problems with the development of a pattern on the coated surface as coat weights increase.

The inclusion of pigmentation in the size press contributes to better brightness, opacity and control of the surface properties important to printing. The nature of the size press is such that the solution applied there will be driven into the web as well as applied to the surface. Therefore, the size press always serves two functions: the increase in bonding strength and the modification of surface properties.

Physical Modification: Calendering

The machine calender is a stack of steel rolls at the dry end of the paper machine through which the web is passed. The steel rolls press the web either through their own weight or through the use of loading cylinders. The pressure on the web causes the surface to be compressed, producing a flatter surface. The smoother surface will print better, will feed better in copy machines where single sheets must be pulled from a pile and will simply have a smoother feel. The calender stack will also reduce the thickness of the paper, and may be used by some people for that reason more than for the purpose of surface modification. The changes made in the web are not permanent; the web will expand and the surface will revert toward its original condition if the web is wetted or even brought into a room with high humidity conditions. However, the web probably will not revert completely to its original condition and will still show an improvement over the noncalendered sheet.

The operation of the calender stack is not as simple as the description above would imply. The steel rolls at the bottom of the stack (and at the top if the stack is externally loaded) will tend to bend due to the weight of the rolls and the pressure used. If these rolls do bend, the web is subjected to more pressure at the edges than at the center. The center will therefore be rougher and not reduced in thickness as much. To counteract the tendency of the bottom (and top) rolls to bend, they are made larger in diameter than the other rolls. They may also be crowned, meaning that the rolls are larger in diameter in the center than at the outside edges. When weight or pressure is applied to the top of the crowned roll, it will bend, causing the bottom to bulge. If the roll is designed properly for the pressure exerted on it, the deflection of the roll will be equal to the crown, and the top of the roll will become perfectly flat. Obviously, only one combination of pressure, roll diameter and crown condition will result in a flat roll surface. If the roll is subjected to a different pressure, the top will not be flat. We have recently seen the development of more complex rolls that have the ability to either control the crown or remain flat under a wide range of pressures. These rolls have contributed to better control of both uniform smoothness and uniform caliper of the web.

Another problem associated with the calender is the development of streaks in the web. If the web comes to the calender stack with either wet streaks or streaks in the machine direction that are of different caliper, the rolls in the stack respond by heating up more or less, depending on the web conditions. The diameter of the rolls in the stack will expand due to heat generation, and may expand enough to create increased pressure in the hot areas. Therefore there must be a provision for controlling the temperature of the rolls in the stack. Early methods consisted of shining hot lights and blowing cold air on the rolls. Better manufacturing methods now give us rolls that are bored so that liquids may be circulated through them to maintain uniform temperatures.

In the combination boxboard machine, liquid solutions can be applied to the surface of the web by the use of devices called *water boxes*. The solutions may be chemicals that will help develop water resistance in the coating, starch solutions applied to one side only to help control curl, or simply water to help develop the gloss of the coated board. The development of smoothness and gloss in the calender is sensitive to the moisture content of the web passing through the stack. The higher the moisture, the higher the gloss. However, the web is normally dry when it reaches the stack.

Some machines have more than one stack. Combination boxboard machines commonly do, simply because it takes that many nips to develop the desired gloss in coated board. A stack is sometimes installed in the middle of a dryer section or just before a size press to help press the paper before it is completely dry, or before the size application to help keep the size on the surface. These stacks are referred to as *breaker stacks* because they break the dryer section into smaller sections.

The yankee dryer described in the preceding chapter functions to dry the web and may also be used to modify the web. Use of the yankee without the creping doctor creates a smoother, flatter web; use of the creping doctor develops a rough, absorbent surface. The web may be embossed by passing it between two rolls with raised patterns, but this again is considered a converting operation, is not commonly done on the machine and will be discussed in Chapter 11.

PIGMENTED COATING

Pigmented coating is actually just another form of surface treatment, but because it is more complex than the other methods of surface modification, it is treated here as a distinct web modification method. As the name implies, a pigment is applied to the surface of the web. The pigment is applied in water, with an adhesive present to hold the pigment on the surface of the web when it is dry. The pigments used are primarily the same three used as fillers: clay, calcium carbonate and titanium dioxide. Since the pigment particles are substantially smaller than fibers, the coating operation creates a surface that is smoother than the uncoated surface and that has a much finer pore structure. These two factors improve the printing characteristics of the web. The coating may also improve the brightness of the web if the pigments are brighter than the fibers. The addition of scattering surfaces in the coating will improve the opacity of the web, unless the web is already opaque. The final benefit to be obtained from pigmented coating is a possible improvement in gloss. The coating itself may or may not improve gloss;

Table 7.1. Pigments for Paper Coating

Grade	Particle size[1]	Brightness	Value or major use
Clay-kaolin (90% of all pigments used)			
Filler clay	40-60%	81-85	Low cost, some bulk
Coating No. 2	80-82%	86-92	Industry benchmark, most used
Coating No. 1	92%	86-92	Improved brightness and gloss
Fine No. 1	95-100%	86-92	Further improved gloss
Delaminated	80-??%	86-92	Ink holdout and smoothness
Sheens	95-??%	86-92	Ink holdout and gloss
Calcium carbonates (second largest—at 6%)			
Precipitated	75%	96	Brightness, ink receptivity
Ground limestone			
Medium	45%	92	High solids, good brightness, low gloss
Fine	75%	95	High solids, brightness, good gloss
Ultrafine	95%	95	High solids, brightness, high gloss
Titanium dioxide	0.25 ave.	96-97	Excellent opacity, high brightness, good gloss
Plastic pigments	0.3 ave.	?High	Good gloss, brightness, bulk

[1] % refers to cumulative percent of particles finer than two microns; titanium dioxide and plastic pigments are expressed as average particle size in microns.

but if an improvement in gloss is desired, the coated paper can be supercalendered to improve gloss.

The operations involved in pigmented coating therefore are: preparation of the coating raw materials, application of the coating to the web, drying of the coated web and gloss development or other handling operations. Before detailing each operation, however, a discussion of the individual raw materials is useful.

Coating Raw Materials

Pigments

As indicated earlier, three pigments—clay, calcium carbonate and titanium dioxide—account for the majority of the coatings applied. Many more pigments can and are being used to small extents by some coaters. Some pigments are so effective at modifying paper properties that only small amounts need to be used. Others are too expensive for the benefit obtained and can't compete with the three major pigments. Even if we restrict our discussion to the basic three, we find that there are many grades of clay and carbonate available and that the selection of which pigment or combination of pigments to use is quite complex.

The different grades of *clay* can be distinguished on the basis of brightness, particle shape, particle size distribution or combinations of these factors. Being naturally occurring pigments, clay particles are found in a wide range of sizes.

The processing of the pigments focuses primarily on fractionating the raw material into grades of different particle size ranges. The coarse-particle grades are used as fillers, and the finer grades are used for coating. The finer the particles, the better the gloss obtainable from the coating. Essentially two brightness ranges are available for most of the particle size ranges. Some of the grades available are listed in Table 7.1; the grade designated No. 2 is the most commonly used. Coatings are occassionally formulated using only clay, but will most commonly be made of a mixture of pigments, with clay as the major ingredient and the others added in small amounts to obtain special properties.

Calcium carbonates are also available in a variety of grades. The major breakdown in carbonate grades is between the natural or ground grades, those made by grinding white limestone, and the grades produced by chemical precipitation. Both grades are brighter than clay and can be used to produce higher brightness papers. The precipitated grades are generally of a finer particle size and can be used to obtain better gloss, smoothness or control of ink absorbency. The ground grades are available in a range of particle size distributions that are at best as fine as the No. 1 clay grades. However, particle size ranges are deceptive, since the finer carbonates can produce gloss values higher than comparable clays. The coarser carbonate grades have long been recognized as being valuable in preventing the development of gloss. Among the more important reasons for the use of the ground carbonates is their ability to be dispersed in water at high solids levels and still have relatively low viscosities. Some coaters are using carbonates to obtain higher coating solids, which reduces the cost of drying the coating after application, as well as potentially improves some properties of the final sheet.

Titanium dioxide, as discussed earlier, is distinguished by its different refractive index and is primarily used for developing opacity. *Plastic pigments* are primarily polystyrene spheres, which were developed originally to allow reduction of coat weight due to their low specific gravity. Plastic pigments are now being used to help develop gloss and brightness and control printing properties. *Satin white*, a white, needle-shaped pigment made by mixing alum and lime, has enjoyed wide acceptance in Europe, but has been in and out of favor in the USA because of a tendency toward high viscosity in the coating dispersion. There are many more white pigments that can be used to improve some specific properties, but are limited in use due to high cost, low availability or adverse rheological properties in the dispersed phase.

Rheology is the study of stress-strain relationships for rigid materials, or is the study of the viscosity behavior of liquids under different shear conditions. The rheology of the coating is important and must be controlled so that the coating can be pumped easily or flow under gravity (low-shear conditions), and later be able to perform well under the high-shear conditions of the coating application system. The subject goes beyond the scope of this book, but is discussed in most references on coating and must be understood by the coater.

Any pigment to be used in coating must be able to be dispersed in water; it is also desirable that it have a fairly low viscosity when dispersed and allow the solids of the coating to be raised as high as possible. Many of the pigments are supplied in a 70% solids by weight slurry or higher, eliminating the need for the coater to disperse the pigments. However, these slurries also limit the maximum solids to which the coating can be raised. Some pigments are therefore purchased in a dry form to allow higher solids or because too little is used to justify buying and storing such large quantities. The dispersion is created by a combination of high-energy mixing and the addition of dispersing agents to stabilize the dispersion.

Adhesives

The adhesive is needed simply to bind the pigment to the surface of the web and to itself. There are many materials that can satisfy this need. Each adhesive will have its own special advantages and disadvantages when compared with the others; there are also similarities between adhesives. The adhesive is used at a relatively low level, generally as low as possible. The final dried coating is not a continuous film, but rather a porous structure of pigment particles cemented together at their contact points by the dry adhesive. The coating, therefore, gets its light-scattering capability from the air pockets trapped in this structure as much as by the individual pigment particles. If we use too much adhesive, it begins to fill these voids and reduce the light-scattering potential. In this respect all adhesives are similar. Adhesives differ with respect to: the amount of adhesive needed to satisfy the strength requirements of the coating, the water resistance of the dried adhesive and how well it allows the development of gloss and other properties.

The amount of adhesive needed is determined by the intended end use of the coated web. Offset printing subjects the coating to a fairly high force, called a *picking force*, when the thin ink film between the blanket and the coated surface is split. Other printing processes will create picking forces, but the offset force is generally considered the highest. Water resistance is needed to varying degrees in coated products to be used in packaging applications, and to a lesser degree so are scuff resistance, lightfastness and many other properties.

The most commonly used adhesives are listed in Table 7.2, where we see that they are divided into two general categories. This division is made because the two categories have similarities beyond the general discussion already presented. The biggest differences are in the properties of the coating before it is applied to the web. The latexes are generally obtained as a 50% solids by weight emulsion, and will allow the coating to be raised to a higher total solids level or will allow lower viscosities at equal solids levels than the solubles. A higher solids level is desirable to reduce the cost of drying excess water from the coating. However, some coaters use soluble adhesives to get some "body" in the coating mixture, to help reduce penetration of the coating into the web. As with pigments, many factors affect the selection of an adhesive for each application. Furthermore, although we can discuss general trends for properties obtainable from each material, absolute statements are few and relationships may change as other parameters are altered.

Starches are the largest category of adhesive, both from the standpoint of being the most used, and also because they are available in a wide range of grades. Although starch can be obtained from many sources, most is made from corn, with a little obtained from potatoes. Natural or pearl cornstarch is too high in viscosity to be used and must therefore be converted in some way. The largest tonnage is converted in the mill, primarily by the use of enzymes that shorten the molecular chains, thereby reducing viscosity. Thermal chemical conversion involves the use of a mild oxidizing agent, which also allows in-mill conversion and viscosity control. Starches can also be purchased from suppliers in a preconverted form, having already been treated to modify the viscosity and usually also the chemistry of the starch molecule. These conversion operations increase the cost of the product, but also enhance its usefulness in certain grades.

All of the starch grades are sold in a dry granule form. The dry granules are then suspended in water, heated to cause the granules to swell, and subjected to shear to break the granules into the individual starch molecules as much as possible. This can be accomplished in batch operations using steam and high-speed mixers, but may cause foam to be generated. Continuous jet cookers mix the starch slurry with steam in a venturi jet, getting simultaneous heating, swelling and shear to break the granules. Jet cookers are very popular and account for production of a large share of the starch used today.

Proteins have been obtained from milk in the form known as casein, but are more commonly obtained from soybeans in today's market. Protein adhesives are a complex mixture of natural materials found in the bean, and may contain some carbohydrate fractions as well. Although similar to starches in their ability to control viscosity of the coating, they can be raised to a higher level of water resistance and therefore have enjoyed wider acceptance in grades where this property is important, primarily packaging grades. Proteins are also sold as dry powders, which must be cooked. Cooking involves raising the temperature to control or lower viscosity, and pH adjustment to help dissolve the protein molecules in the adhesive. It must be noted that these so-called soluble adhesives are actually macromolecular suspensions and are not true solutions.

The *latex emulsions* as a group can give the final coating the highest level of water resistance. Gloss and brightness of coatings containing latexes are sometimes considered better than those possible with the soluble adhesives, but other factors can negate this statement in some situations. The three major latex groups are available in several different forms, limiting the comparative statements that can be made. *Styrene/butadiene* is the widest used and generally is considered good for gloss, strength and general usefulness. *Polyvinyl acetates* have advantages in some packaging applications where glueability, lightfastness, low residual odor and porosity of the coating are needed. *Acrylics* also are excellent in these packaging applications, but have suffered price problems and are not as widely used.

Coating Preparation

The coating mill will have a separate department or area, sometimes called the *coating kitchen*, where the coating is prepared. Each mill must develop its own system and assemblage of equipment based on the grades to be produced and the types and complexity of raw materials to be processed. Pigment dispersions must

Table 7.2. Adhesives for Coating

	Properties					Use	
Adhesive	Strength	Water resistance	Gloss development	Ink receptivity	Special characteristics	Markets	Volumes
Soluble							
Starch							
Enzyme	Good	Poor	Fair	High	Inexpensive	Publication grades	High
Thermal	Better	Poor	Fair	High	Moderate cost		
Ethylated	Better	Fair	Fair	High	Improved performance		Moderate
Proteins	Better	Good	Fair	Medium	Glueability	Packaging	Low
Latex							
Styrene/ Butadiene	Best	Best	Best	Low	Good all-around performance	Printing papers	High
Polyvinyl acetate	Better	Best	Best	Medium	Glueability	Packaging	Medium
Acrylics	Best	Best	Best	Medium	Lightfastness	Packaging	Low

be prepared, unless they are purchased in the slurry form. Likewise, if soluble adhesives are to be used, there must be equipment for their preparation. Once ready for use, the two must be metered together, blended, screened, stored and pumped to the application system. The simplest system could be tanks to receive and store slurry pigments and latex emulsions; the more complex systems need adhesive cooking equipment and perhaps several tanks for pigment dispersion, as well as storage tanks for these ingredients after they are prepared.

The nature or composition of the coating varies over a wide range and is affected by the grades being produced and the method of application. Coatings are aqueous dispersions ranging in total solids from about 50% to more than 70%. The ratio of adhesive to pigment also varies from as low as 6 to as high as 25 parts dry adhesive per 100 parts dry pigment. The adhesive level is determined by pick resistance and water resistance or other forms of strength needed in packaging applications. The adhesive portion may be only one component, but in the majority of cases is a blend of one soluble adhesive and one latex. The combination gives both the wet and dry coating characteristics desired. In most cases the selection involves a compromise between demands. The pigment portion will also tend to be a mixture designed primarily for its dry coating properties, but perhaps compromised by cost or wet rheological considerations.

Coating Application Methods

The coating method used is largely dependent on the grade of paper or paperboard being produced, primarily from the basis of the coat weight desired. The speed at which the machine must be run to be economically competitive is also important, as is the effect the application system may have on the surface of the coating. The coating operation may be performed on the paper machine with the coater being an integral part of the machine, or it may be an off-machine or conversion type operation.

All application systems need to perform three related functions: (1) the coating must be applied uniformly to the entire surface of the web, leaving no uncoated areas or skips; (2) the amount of coating on the surface must be metered to ensure that it is of the desired thickness or coat weight over the entire web surface; and (3) the surface should be made as smooth and uniform as possible. We will see that some coaters are designed to combine all of these functions, some seem to disregard or fail to consider some of these points and others have special consideration for each function.

Size Press

The size press can be used for coating by the addition of pigments to the starch normally applied. The formulations will seldom exceed 50% total solids, and will require higher binder levels than normal coatings, since the binder still will penetrate into the web. A variety of pigments and binders are used, including latex binders, to obtain the desired properties. Coatings vary from pre-coats over which coatings will be applied by other methods, to lightweight coated paper, which is ready to be used for newspaper inserts or low-quality magazines.

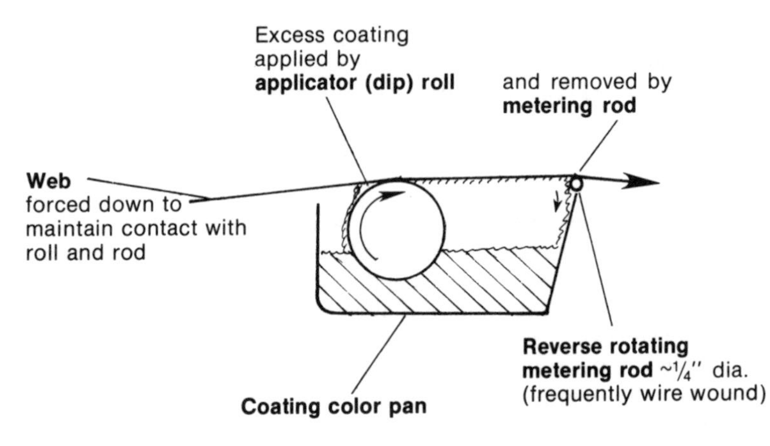

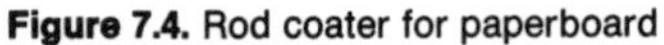

Figure 7.4. Rod coater for paperboard

Air Knife Coater

Another coater that has had wide application in the coating of paperboard is the air knife coater, shown in Figure 7.5. This coater applies an excess of coating with the dip roll and uses a narrow wedge of high-velocity air to meter and smooth the surface. This coater is capable of applying heavy coat weights of up to 30 lb/side (3,300 ft^2 basis, about 45 g/m^2), but is limited in speed to 2,000 fpm (10 m/s) or less. Higher speeds are possible on machines that include a smearing or smoothing roll between the applicator and the air knife. The lower speeds and heavy coat weight capability are well suited to use on the cylinder machines used to make combination boxboard. These coaters also apply a very uniform coating

Figure 7.5. Air knife coater

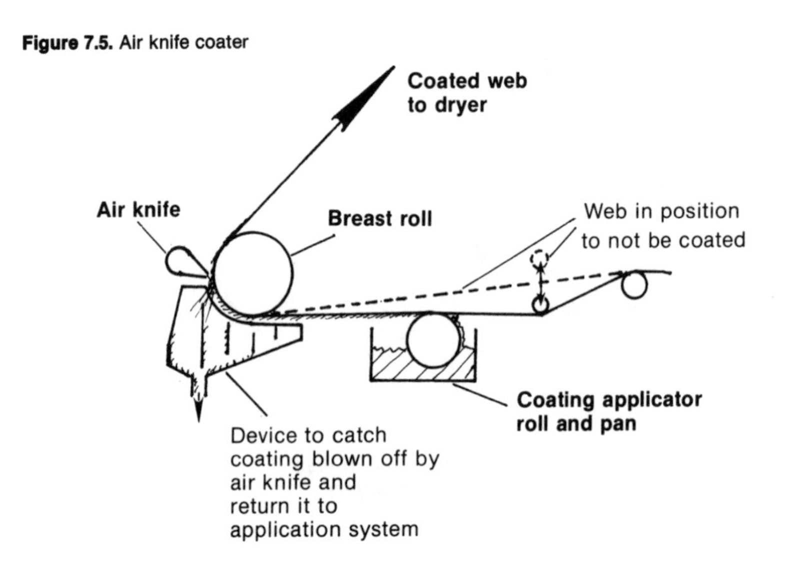

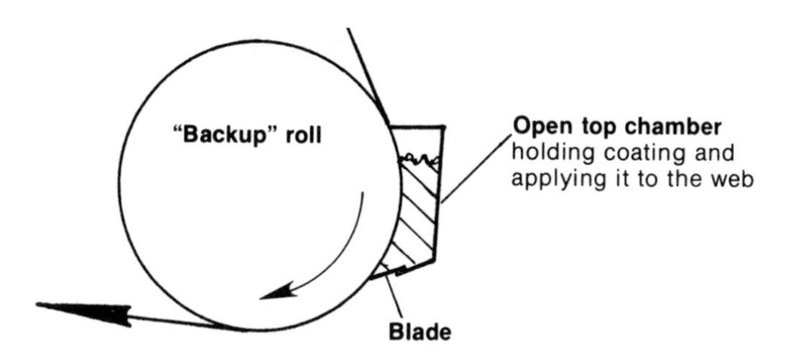

Figure 7.6. Pond or puddle type blade coater

thickness and are described as contour coaters since the coating follows the surface of the base web. This characteristic has appeal for offset printing grades, explaining the use of these coaters for off-machine-coated merchant grades of paper. Air knife coaters are still in use, but their use is not growing as rapidly as that of blade coaters because their lower machine speeds and off-machine operation are less economical than the newer blade coating methods.

Blade Coater

Since their introduction in the 1950s, blade coaters have gone through a tremendous transition. Among the early designs were the *pond or puddle type coater* shown in Figure 7.6. The coating is held in the pond as the paper passes through it and out under the blade, where the coating is metered and smoothed. The blade is a flexible steel strip about 0.012 in. thick, 3 in. wide and long enough to reach across the width of the web. These coaters are plagued by problems with sealing the coating in the pond at the outside ends and around the blade, splashing of the coating during operation and a severe mess when the web breaks and the pond must be dumped as the blade is pulled away from the backing roll.

In newer blade coater models, the blade is separate from the applicator, leading to their description as *trailing blade coaters.* The drawing in Figure 7.7 is of one such coater, using a roll applicator and what has come to be known as a *beveled blade.* The blade is held by one edge and pushed against the surface of the web by an air-loaded tube on the back side of the blade. The pressure on the web surface is a combination of this loading pressure and the stiffness of the blade. The tip of the blade is beveled to the same angle as the angle between the blade and the web so that the flat part of the tip rides on a thin film of wet coating between the web and the blade. This coater is capable of very high speeds (5,000 fpm; 25 m/s) and a range of coat weights from 6 to 15 lb/side (3,300 ft^2 basis, 9 to 20 g/m^2). The coating is quite flat on the surface but may vary in thickness if the rawstock (uncoated) web is rough. This type of coater is very versatile and is used on most grades of paper and paperboard. Any of the beveled blade type coaters are susceptible to problems with scratches caused by particles trapped under the blade or a breakdown of the coating caused by the high shear conditions found between the blade and the web.

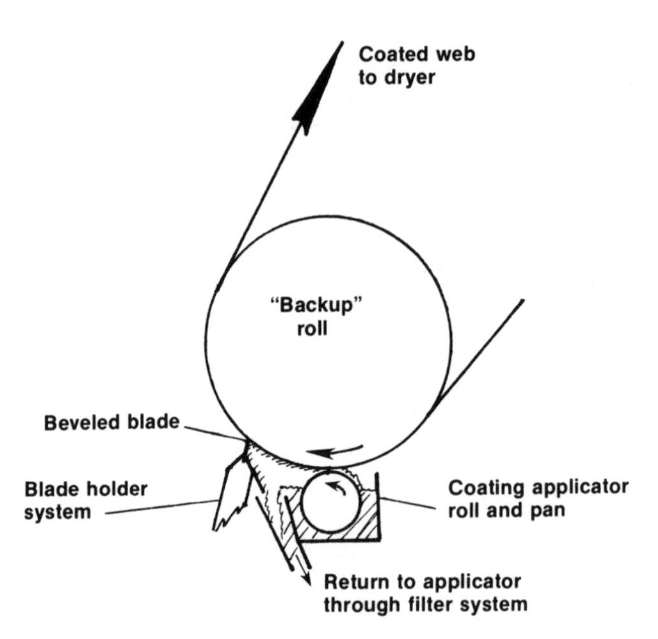

Figure 7.7. Blade coater: Beveled blade with roll applicator

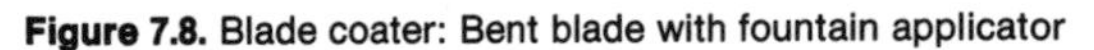

Figure 7.8. Blade coater: Bent blade with fountain applicator

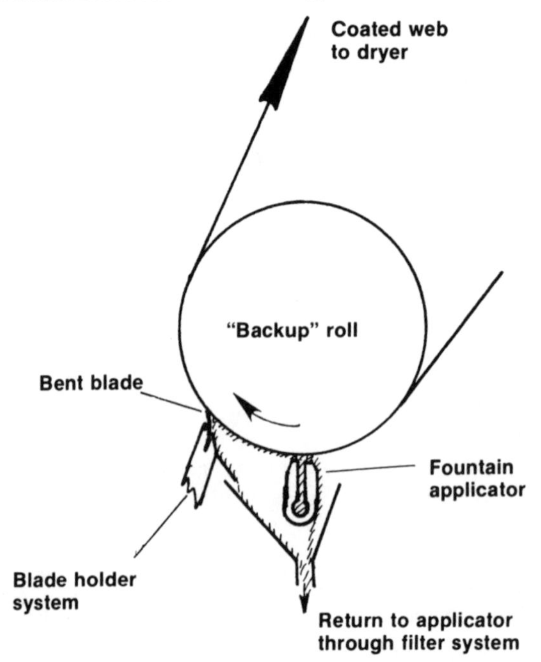

Another modification in application methods is shown as the *fountain application* in Figure 7.8, along with a different concept in blade metering and smoothing methods. The applicator systems and blade concepts are interchangeable and do not need to be used in the combinations shown here and in Figure 7.7. The fountain applicator has a narrow slot at the top where the coating is applied to the web. The coating is applied with a lower pressure than the roll applicator, which is intended to reduce the potential for pushing the coating into the web. The blade shown in this drawing is different from the preceding design in that the coating is smoothed and metered by the face, rather than the tip, of the blade. The blade is still held by one edge, but is flexed more and the pressure tube may be brought all the way up to the tip of the blade. This *bent blade* allows higher coat weights, is more of a contour coater and is less prone to scratching than the beveled blade. A comparison of blade angles, types and their operating characteristics is given in Figure 7.9. The rod tipped blade has been mentioned already for use in the size press replacements, but is also used in regular coaters. The rod allows higher coat weights than the beveled blade, gives a more contour coating and does run with fewer scratches. It is being used primarily with board coatings, but is also used on some paper applications. The beveled blade is primarily used on high-speed light-weight coaters.

The most dramatic recent design is typified by the short dwell coater shown in Figure 7.10. This coater removes the roll or fountain applicator and applies the coating immediately before the blade. In this design the pressure pulse of the roll applicator is removed along with the pressure-induced penetration of the coating into the web. This coater has seen its widest application for light-weight or publication coatings, where the penetration of the coating into or through the web is most detrimental. The penetration can cause the backing roll to become covered with coating; or pinholes in the paper can allow the coating to glue the web to the backing roll, causing web breaks. The advantage of this coater is

Figure 7.9. Blade designs

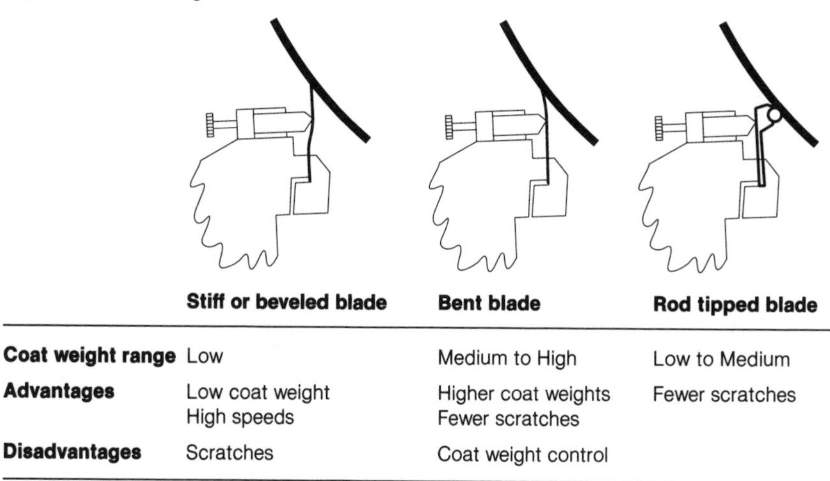

	Stiff or beveled blade	**Bent blade**	**Rod tipped blade**
Coat weight range	Low	Medium to High	Low to Medium
Advantages	Low coat weight High speeds	Higher coat weights Fewer scratches	Fewer scratches
Disadvantages	Scratches	Coat weight control	

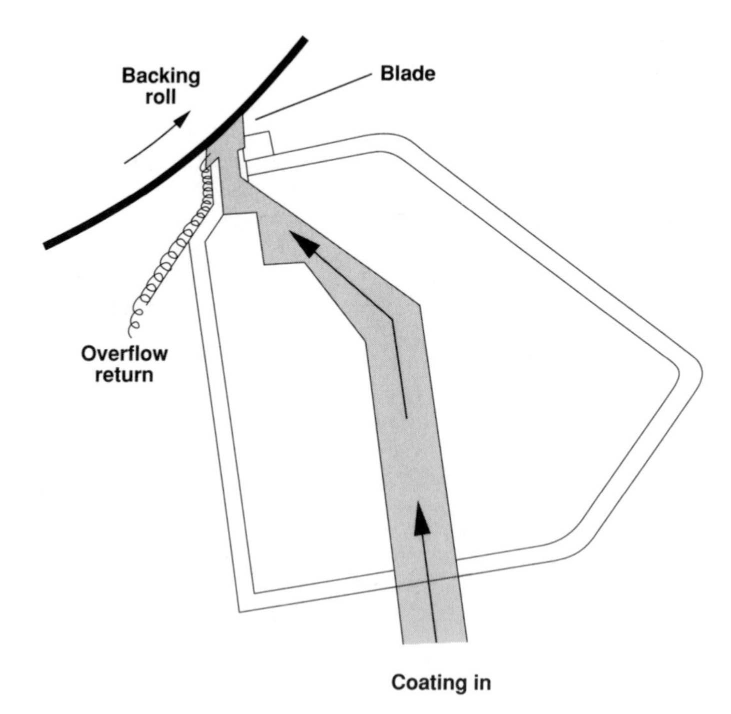

Figure 7.10. Short dwell coater

Off-machine coater at Bowater's Catawba, South Carolina, mill showing unwind in the foreground and tunnel dryers above the first blade applicator in front of the operator.

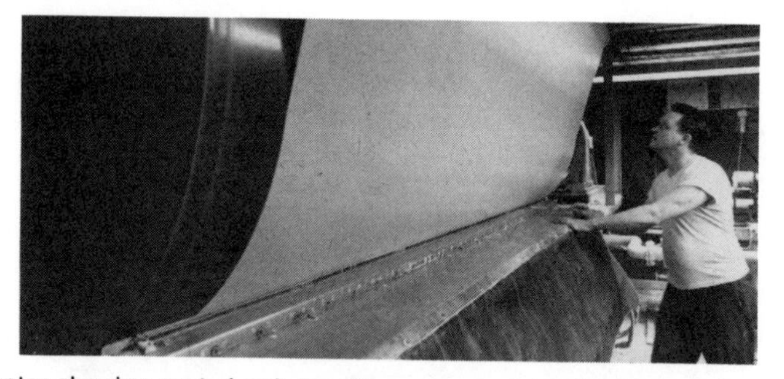

Blade coater showing coated web traveling up away from the blade.

therefore in improved operating efficiency, not final sheet quality. In fact, it has been shown that with high wood content papers, the base sheet swells after the coating has been applied leading to lower smoothness than other application methods.

Coating Drying

The coating is dried primarily with hot air. Some use of infrared radiation to heat the web can be found, but not without hot air to finish the job. Dryer cans are also used, but only after the coating has been dried enough so that it will not stick to the surface of the cans. With coaters that apply coating to only one side of the web, the web can be wrapped around a drum with the wet side out, and the air blown on the coated surface. This type of dryer is called a *hot air cap or hood* and is shown in Figure 7.11.

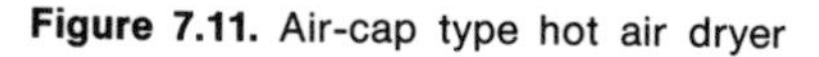

Figure 7.11. Air-cap type hot air dryer

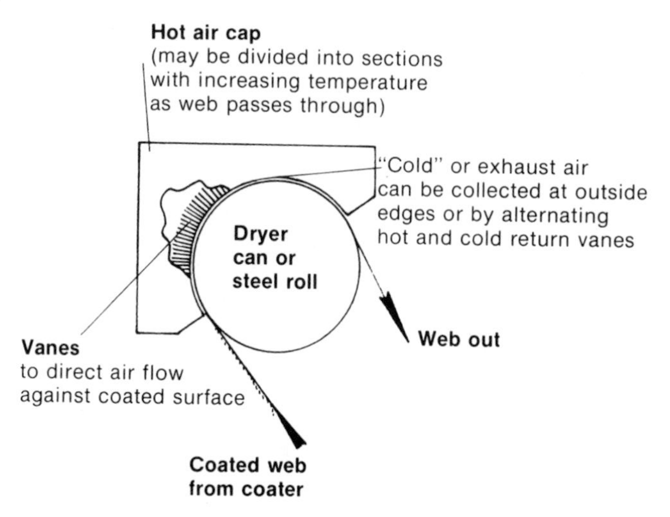

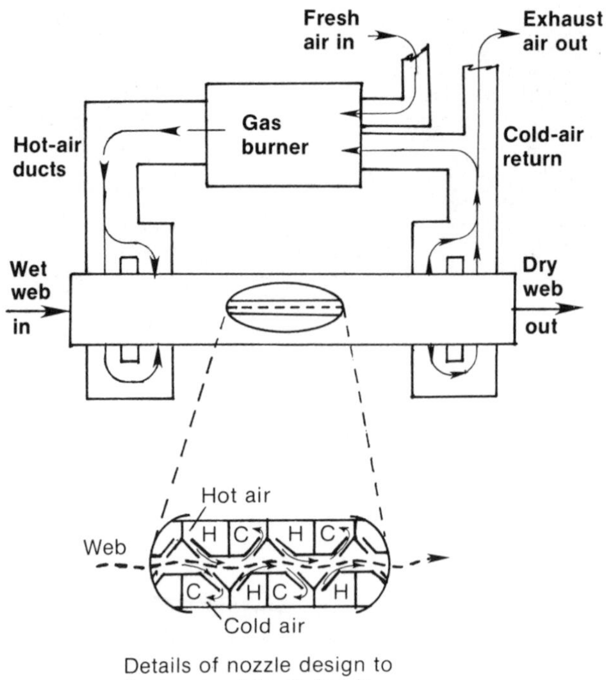

Figure 7.12. Tunnel or "floater" dryer

Coated paper is frequently dried by passing the web through a tunnel and blowing hot air on the surface. The tunnels may be designed with jets to cause the hot air to impinge on the coating surface, with the web supported by carrier rolls.

If both sides are coated simultaneously, the web must be sent to a special dryer tunnel, which supports the web with air. The *floater type dryer tunnel* shown in Figure 7.12 is designed for this purpose. Floater type tunnel dryers are also used for webs coated on only one side, since they have been shown to be very efficient.

Figure 7.13. Coater design

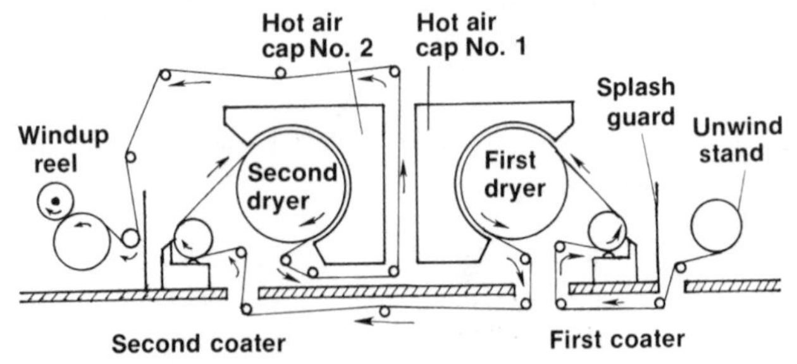

Overall Coater Design

The first consideration in coater design is generally whether the coater is to be installed as an on- or off-machine operation. The question is primarily one of economics. The relative speeds of the coater and the paper or paperboard machine, the drying capacity of the machine and the physical space available all influence the decision. Equally important are whether both sides need to be coated and whether it is desirable to apply more than one layer of coating to one or both sides of the web. Combination boxboard and label papers need coating only on one side, but may receive multiple coats on that side. Newsprint and low-cost publication grades need to be coated on both sides, but will not demand a high enough price to allow double coating. Some high-priced specialty grades may even be triple coated on both sides.

Many of the design considerations can be demonstrated in one drawing of an off-machine blade coater designed to apply one coat to both sides (Figure 7.13). The web in this drawing is coming from a flying splice unwind stand so that the coater can run continually, without the need to stop it to change rolls on the feed end. If the coater were to be installed on the machine, the unwind would be eliminated and the web would be coming from a dryer section of a paper or paperboard machine. Most coaters apply the coating to the bottom side of the web and then raise the web up and turn it over so the wet coating is on top, as shown in Figure 7.13. In order for both sides to be coated in this manner, the web must be fed first in one horizontal direction and then in the other. The selection of which side to coat first is not totally arbitrary, but is based on where the web will go after drying and how difficult it will be to feed to the second coater, as well as on the properties of the web.

In Figure 7.13 it can be seen that the web has been passed beneath the first coater head to apply the coating first to the felt or top side of the web. Passing the web under the floor protects it from the possibility of being splashed with any spilled coating. After the first side has been dried, the web is passed through the second coater head and dried. The coating machine shown is equipped with a drum winder typical of most paper and paperboard machines. If the coating were done on the machine, this part of the coater would be quite the same. If more than one coat were to be applied to either or both sides of the web, this design could be expanded to include as many heads as desired. Early air knife and some blade coaters were designed to coat only one side of the web. If the web required coating on both sides, the rolls would have to be passed through the machine twice to be double coated. Because such designs are obviously less efficient than multiple coaters, they are no longer being built new and are being phased out in most mills.

Supercalendering and Gloss Development

The coated web will be smoother than the uncoated web, but may not show any improvement in gloss. The coated and uncalendered web is referred to as a *matte coated grade*, is generally low in gloss and may be used for low-cost publication or book grades. These grades are less costly to make, and the low gloss is desirable in some applications; however, the surface is not as resistant to abrasion and may cause dusting or piling problems when printed. By giving the web a low degree of the right kind of finishing, a little gloss can be developed while the surface is flattened and toughened. These grades are designated *dull finish*. By using supercalendering or other gloss-developing methods, high-gloss grades (known as *enamels* or simply *high-gloss grades*) can be produced.

A machine calender with all-steel rolls, as described in Chapter 6, may be satisfactory for combination paperboard or low-cost publication grades, but will not allow the development of the high gloss of enamels. The supercalender is used for high-gloss coated paper. The supercalender is shown in Figure 7.14 to have a stack of rolls alternating between steel and filled rolls. The latter are filled with either cotton or plastic materials, which makes them slightly softer than the steel. As the web passes between these rolls, the filled roll is dented or deflects, causing a slight amount of creep or slipping in the nip. This, combined with the high pressures used in the supercalender, creates a polishing effect on the surface of the web. Since the side in contact with the steel roll is polished more than the other, the stack has two filled rolls together in the middle so that both sides will be subjected to this polishing effect as the web passes through.

A higher level of gloss is obtained by a process known as *cast* or *chrome coating*. In this process the wet coating is applied to one side of the web and then pressed against a chrome-coated dryer can. The coating is dried in contact with the chromed surface, thereby picking up a flat surface reflective of the chromed

Figure 7.14. Supercalender design

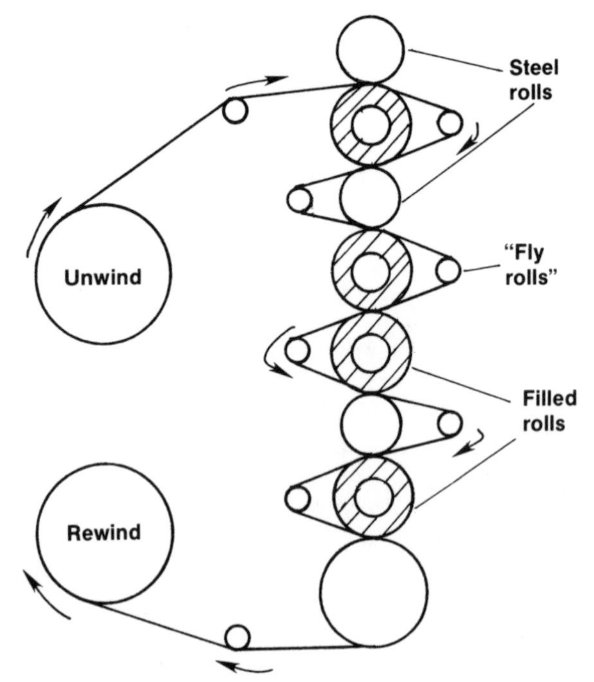

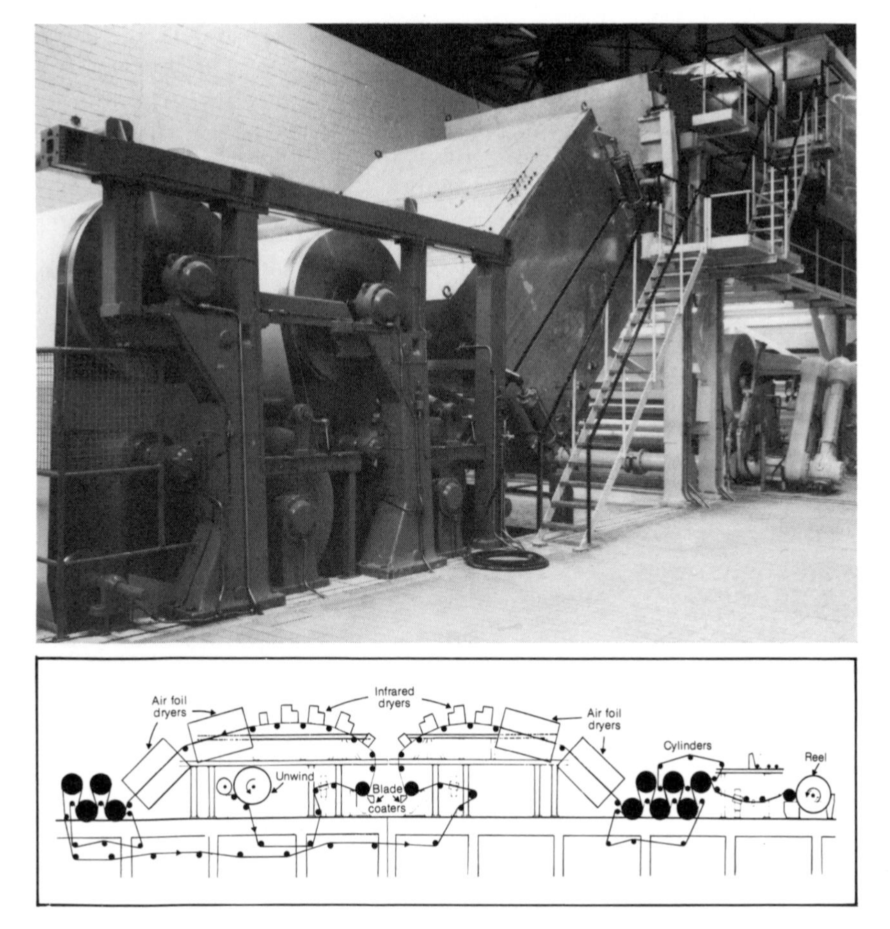

An off-machine coater designed to blade-coat both sides of the web.

surface. This process is slower and more expensive to operate than the other processes and is restricted to high-cost applications in advertising or packaging.

A gloss-developing technique similar in concept to the chrome coater is currently becoming quite popular. In this device, known as a *gloss calender*, the coating is pressed against a polished, hot surface while still damp, allowing it to dry and obtain some of the flat, shiny surface characteristics of the polished surface.

All of these finishing techniques give additional variability and choice to both manufacturer and purchaser of coated grades. With all of the choices in raw materials, coating methods and finishing techniques, it is no wonder that no two mills are exactly the same. All of these choices will influence both the desirable and undesirable qualities of the final product. In all cases there are compromises to be made between cost and quality as well as between competing properties of

end of the press. Paperboard has a higher tolerance for water and can be printed with water-based flexo with minimal drying.

A third, rapidly growing application for flexo printing is the printing of pressure-sensitive labels. In this application the printing is done on the combined label, pressure-sensitive adhesive and release-coated backing paper. The use of the soft plate helps protect the label just as it helps protect the film or corrugated board in those applications. The ink used for labels needs to be dried quickly but cannot rely on penetration of water into the label, as that could cause it to curl. The recent development of inks that cure by exposure to ultraviolet radiation has been applied to this use with a great deal of success; in fact, these inks have been key to the rapid growth enjoyed by this segment of the industry.

The actual press designs are very specific to each application, and are not easy to generalize. Most are simply the installation of a plate and inking system, similar to the one shown in Figure 8.3, that can print one side of the web. Drying depends on the type of application and ranges from none to a long tunnel to blow hot air on the web.

OFFSET PRINTING

Plate Design and Ink Characteristics

The name *offset* comes from the fact that the ink is not transferred directly from the plate to the paper, but is transferred by a rubber-covered mat called a *blanket* on an intermediate cylinder. The process is more accurately called *lithography* or *offset lithography*.

The basics of the process are that the plate has a planar or flat surface, and the ink adheres only to certain portions of the plate, while water adheres to others. The plate is a metal sheet that has been coated with a special light-sensitive material. One method of platemaking involves exposing the plate to light and then washing away the coating in the nonimage area. The remaining coating will reject water and be oil or ink receptive. The exposed metal surface will accept water and, when covered with a thin film of water, will reject oil or ink. The press therefore must be set up to apply both water and ink to the plate constantly. The fact that the press uses water makes it more difficult for the press operator to control, and also leads to potential problems when printing on paper. The use of the offset blanket, which contacts the entire surface of the sheet or web, whether printed or not, leads to another set of potential problems. The paper or board being printed must be clean and able to resist the tack or *picking* (upward pulling force) of the blanket and ink, since any material from the substrate that lodges on the blanket can interfere with the print quality. The use of the blanket does reduce the need for a smooth printing surface and does allow this process to produce a better print on rougher paper.

The inks used include nondrying types used on news presses; oxidizing types for sheet-fed presses, which require up to 24 hours to dry; heat-set types for web-fed presses; and ultraviolet-set or cure types used on both web and sheet-feed presses. All offset inks are of the high-viscosity, oily type. Use of the offset blanket allows this process to employ the thinnest ink film, which when combined with use of the blanket creates the highest picking forces of all the common printing processes. The thin ink film can also make this process the most sensitive to the ink holdout characteristics of the web.

Offset Press Designs

Sheet-Fed Litho Press

The oldest of the offset presses are the ones designed to print individual sheets of paper or paperboard. These presses are decreasing in popularity with the development of the web-fed press, but still retain their usefulness in certain applications. These presses are well suited to shops that need to print a wide variety of sheet sizes, since the presses can print on sheets of paper smaller than the maximum size for which they were designed. It is possible to expose only part of the plate, and to feed paper through the press to be printed by the portion of the plate that has been exposed. Web presses must use the entire plate surface since paper is coming through in a continuous web, and any portions not printed would have to be cut out and thrown away. This factor gives sheet-fed presses greater flexibility in handling odd-sized jobs than web-fed presses. Small sheet-fed litho presses are also very popular for office copying or fast printing of letters and other small jobs, although xerographic copiers are also very popular for these applications.

A representation of a sheet-fed litho press is shown in Figure 8.4. This drawing is a two-color press since it has two printing stations. The paper is loaded into the press by hand or on skids, is lifted from the pile by small vacuum suckers and is then delivered to the inclined board and fed into the press. A more recent development is for rolls of paper to be unwound and cut into sheets on the feed end of the press. The sheets are fed into small metal fingers (called grippers) on the impression cylinder, which carry the sheet past the printing blanket. Ink and water are applied to the plate at each printing station, transferred to the blanket and then to the sheet. If the press is intended to apply more than one color of ink, the sheet is next carried by the grippers of a transfer roll to another impression cylinder. This process is repeated until the sheet is ready to be carried to the delivery pile by grippers attached to chain carriers. Presses normally come with one, two, four or more printing stations. The number of stations reflects the type of printing the shop intends to do. Four-color process work is most economically accomplished on a four-color press, but four-color jobs can be passed through a two-color press twice to be fully printed. Presses with more than four colors are used for applications where special colors of ink are required above and beyond the basic four, or where a varnish is applied over the ink to protect it and give the product scuff resistance. If both sides of the sheet must be printed, the sheets must be processed twice as many times, since these presses commonly print only one side at a time. Some presses, called *perfecting presses*, have special transfer rolls that turn the sheet over and allow both sides to be printed in one pass.

Sheet-fed litho presses have traditionally been used for only the highest quality advertising or magazines, which does not necessarily mean that they can always produce better quality work than the other processes. Sheet-fed litho can produce

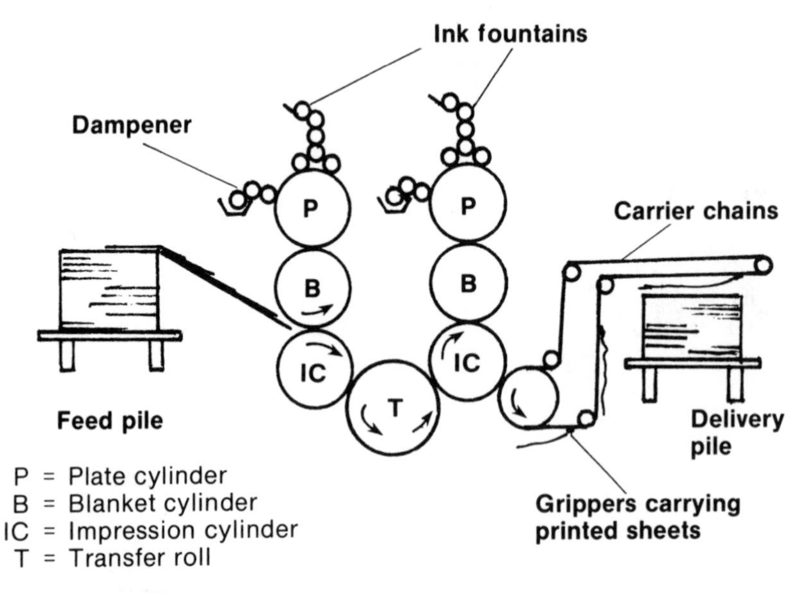

Figure 8.4. Sheet-fed offset press design

excellent quality printing, but is also slow, generates a higher ratio of waste and requires greater operator skill to deliver high quality. Sheet-fed litho may be the most versatile with respect to sheet sizes, surface roughness and size of job, but the selection of which printing process to use is a complex equation in which operating cost, efficiency and speed are as important as quality.

Web Offset Presses

Business Forms Press. Small presses capable of handling several webs of widths up to 22 in. and using either an offset or a direct litho printing process are very popular for the printing of *business forms* such as high-speed computer printout paper, forms for order and receipt machines and similar uses. Each press is different, and therefore this type is difficult to represent in one drawing. Most are characterized by an unwind stand capable of holding more rolls than the maximum number or webs to be found in the final printed and folded form. The press is usually run more or less constantly, with new rolls being placed on the empty unwind shafts while the press is running, so that the new rolls can replace expiring webs with little or no downtime. Webs are generally printed using presses similar to the one shown in Figure 8.4 with the sheet handling equipment and transfer roll removed. This press can only print one side, which is usually adequate for business forms, but both sides can be printed if the sheet is inverted. Many of these presses are also equipped with punches to perforate the webs for the tractor feeders found in high-speed printers or office machines, diecutters to perforate the forms so that individual sheets may be torn off, folders, and devices for loading the finished forms into boxes for shipment. When more than one web is being printed for multiple forms, the punches, perforators, and other accessories may operate on the combined webs after printing.

The use of offset in these operations stems from its superior ability to print lines or other forms of ruling, the relative insensitivity of the process to web smoothness, the relatively low cost of plates and platemaking as well as the ability to print carbonless copy without excess marking.

Inverting the web requires either running the web back in the other direction or using a web inverting cage similar to the one shown in Figure 8.5. The web is first passed around a nonrotating rod or bar mounted at a 45-degree angle to the web travel, which turns the web by 90 degrees and inverts it. To return the web to its original direction, it must be passed around a turning roll and another bar. Each turn inverts the web, and an odd number of turns leaves the web inverted. The position of the last turning bar has been drawn to the side to make the illustration easier to understand. In actual use it would most likely be located directly over the first turning bar. Nonrotating bars are needed for the 90-degree turns so that the web will not be driven to one side as it would be if rotating rods were used. However, nonrotating bars can cause problems if the ink is not set enough or if the paper tends to dust. Loose fibers or coating pigment can be rubbed off by these bars, causing severe problems.

In order for the web to be inverted and folded into booklets or signatures on the end of the press, the press must use a heat-set or quick-drying ink. The ink will be of fairly high viscosity, but will generally be lower in viscosity than the

Figure 8.5. Web inverter

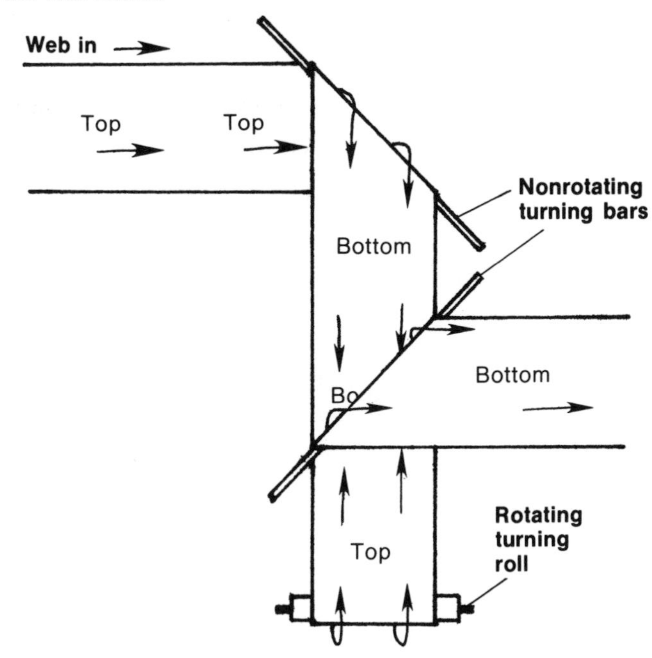

sheet-fed inks. The ink vehicle will contain a volatile fraction that can be rapidly evaporated in the dryers to increase the solids of the ink and to set it so that the web can be inverted or folded. The ink may still have to be dried by oxidation, but it can be set enough to be processed or rewound.

Blanket-to-Blanket or Perfecting Press. The fastest growing web offset press in the USA is the *blanket-to-blanket or perfecting press.* Widely used for publications and direct mail advertising, these presses come in a variety of widths for jobs of different sizes and are usually combined with folder-sheeters, as shown in Figure 8.6. In some of the direct mail uses these presses may include gluing, diecutting, combining with other webs and other conversion operations so that rolls of paper are fed to the press and finished products come out the other end.

As indicated in Figure 8.6, the printing units are the same as in the sheet-fed press with the exception of the open area on the cylinders found on the sheet-fed press. On web-fed presses, the plates and blankets are thin sheets that are wrapped around the cylinders and attached by inserting the ends in a slot in the cylinder. The web-fed press must be able to print all of the web as it passes through and therefore must have nearly solid plate and blanket cylinders. The impression cylinder of this press is another blanket that is printing on the back side of the web, as shown in Figure 8.6. The act of applying ink to both sides simultaneously creates pick forces in both directions and has led to some problems with sheet delamination. This situation also contributes to the *web offset blister* problem. As the web leaves the printing stations, it is passed through a hot oven to set the inks before the web goes to the folder. The web can blister due to the sudden evaporation of water if: the web has too much moisture, the coating is too dense, too much ink is applied to both sides or the web is weakened by the printing stations. Use of inks that set at lower temperatures has reduced this problem, but the potential for difficulty remains.

The press is popular because of its versatility. It is possible to use one web and print four colors on both sides of that web, or feed two webs through the same printing stations, applying less than four colors to each web. If each plate cylinder in Figure 8.6 is large enough to print eight 8½ x 11-in. pages with each revolution, a single web could be printed and folded into a 16-page booklet (called a

Figure 8.6. Blanket-to-blanket or perfecting web offset

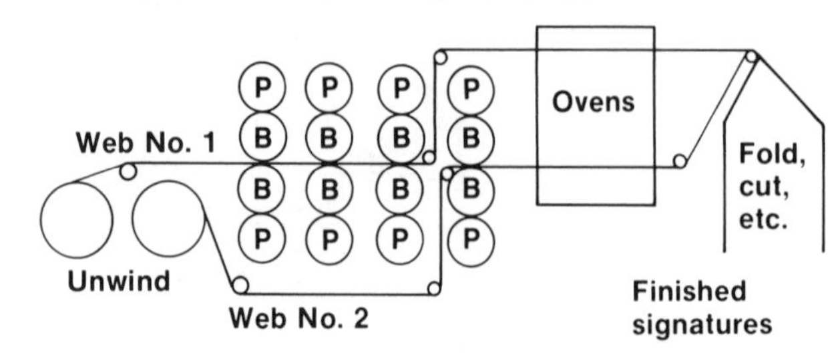

P = Plate cylinder
B = Blanket cylinder

signature) with four-color printing on both sides of all pages. The signatures may be a finished booklet or advertising piece, or they may be sections of a book, which will be combined and bound later. If we use two webs (as shown in Figure 8.6), the size of the signature can be doubled, but the contents cannot all be printed in four colors. The press can be expanded and made even more versatile by the addition of more printing stations on the other side of the folder.

Because of the blanket, these presses are more tolerant of rough paper than letterpress or gravure. They have the same problems as sheet litho with the water and ink balance and the sensitivity of the web to water, but the handling problems are different. Since the web is under tension, the web press can avoid some curl problems, but is subjected to a different set of wrinkle problems.

Common Impression Cylinder Press. The *common impression cylinder offset press* shown in Figure 8.7 combines characteristics of the letterpress press in Figure 8.2. These presses are also referred to as *planetary presses* because of the location of the printing plates around the central cylinder. The plate-and-blanket combination of the litho process is used, as are the same inks and water systems of the other web offset presses. This press lacks some of the versatility of the blanket-to-blanket press since it is not capable of handling more than one web. However, control of the web and virtual elimination of blister and delamination problems are adequate compensation for users who don't mind the limitation in the signatures obtainable. Improved control of the web comes from the fact that the web wraps the impression cylinder between blankets, and is therefore less likely to wrinkle at that point. The reduction in the tendency to blister is due to the fact that only one side is printed at a time, and the web is dried between printing of the first and the second sides. These factors reduce the possibility of ink trapping water in the web and eliminate the simultaneous picking forces on both sides.

The press shown here is designed to print both sides and would be the type used for publication or book printing. Another version, designed to print only one side, has recently been introduced to print folding carton stock.

Figure 8.7. Common impression cylinder offset press

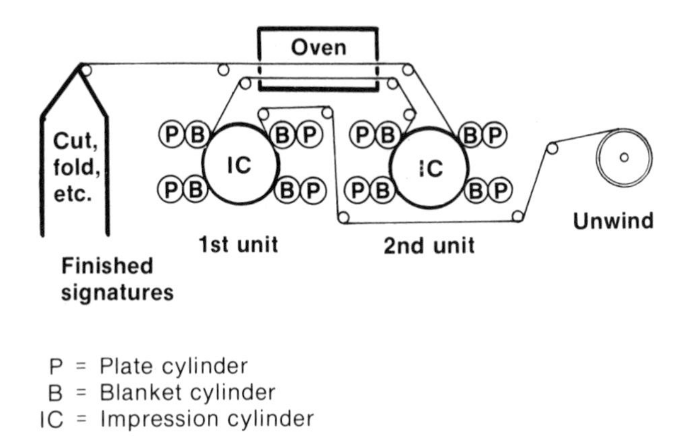

Table 8.2. Printing Grades, Processes and Operations

Grade	Printing process	Papermaking operation(s)	Web modification(s)
Newsprint	Web offset and letterpress; increasing flexo decreasing letterpress	Fourdrinier and twin-wire machines	Seldom more than surface sizing and machine calendering
Uncoated			
Groundwood-containing	Web offset and gravure; some letterpress	Fourdrinier and twin-wire machines; some filler	Surface sizing and machine calendering; some supercalendered
Woodfree book paper	Web offset and gravure	Mostly fourdriniers; some twin wires; extensive use of fillers	Surface sizing and machine calendering
Coated printing grades			
Lightweight coated (LWC)	Gravure, web offset and letterpress	Fourdriniers	Pigment coated by blade or size press
Medium weights (MWC)	Web offset, gravure, some letterpress	Fourdriniers; some use of fillers	Pigment coated by blade and/or size press; more likely to be double coated
Heavy weight (HWC)	Web offset or gravure	Fourdriniers; considerable use of fillers	Pigment coated by blade primarily; very likely to be double coated
Merchant grades	Web offset, sheet-fed litho, some gravure	Fourdriniers; considerable use of fillers	Pigment coated by blade or possibly air knife; almost all multiple coating applications

Uncoated Grades

Uncoated grades include a few more grades than indicated in either of the tables. In Table 8.2 they are divided into woodfree and groundwood-containing, because there are greater similarities in manufacturing and printing. Groundwood-containing grades are used for newspaper supplements and directories (web offset, gravure and flexo), some magazines and books (web offset, flexo and letterpress), business forms (offset) and some miscellaneous products. Woodfree uncoated grades are more likely to be used in better quality books (offset and some letterpress), bond and other business-related grades (offset) and some advertising and merchant grades (sheet litho and web offset). As the quality increases, the use of fillers increases. Many highly filled supercalendered grades are competing successfully with the lightweight coated grades.

Coated Grades

The coated grades category is very large and can be divided in a variety of ways depending on the market, manufacturer, or whoever else might be considered. The breakdown by weight shown in Table 8.2 follows the raw material and manufacturing operations rather than the end use. The lighter weights are most likely to contain groundwood; as the weight goes up, the probability that the grade contains groundwood decreases until we reach the merchant grades, where groundwood is seldom used. Almost all of these grades are still made on fourdrinier type machines, but new installations occasionally use twin-wire machines. The light and medium weights are definitely used primarily for publications, and are also called publication grades. The size of the publication (number of copies printed) as well as the per-copy cost determine the printing process. Gravure is favored for larger runs and web offset for shorter. Letterpress is losing ground to both, but is still a factor.

As we proceed to heavier weights, the cost of the paper increases, as does the quality, and the probability of the paper being used in advertising pieces increases. Again the printing process is tied to the length of the run, with gravure and web offset being the most used. The heavier weights are almost all made on fourdrinier machines, with little penetration by twin-wire machines. *Merchant grades* get their name from the fact that they are sold by the mill to a paper merchant, who warehouses them and then sells them to the printer, usually in smaller lots. These grades were traditionally sold in sheets and therefore printed by sheet litho. However, the increasing use of web offset is causing merchants to handle rolls as well. As we increase the coat weight, we see an increasing probability that coating will be done off the machine and that there will be multiple coating applications. All coated grades may be supercalendered for high gloss or dull finish or sold uncalendered as matte finish.

Special Properties and Tests

Smoothness

One of the most difficult properties for which to test is *smoothness*. Each of the printing processes treats the paper differently and therefore has a different set of demands. To develop a single test to satisfy all processes has long been an elusive goal. The most common type of tester is the *air-leak type*, such as the Sheffield or Parker. These devices press a pair of metal rings against the surface of the web and measure the rate of flow of air between the rings and the surface. Although they may be helpful and can be correlated with some degree of confidence, they cannot be expected to correlate with all processes. Other successful methods are based on proof-printing the paper. There are at least two gravure printers; most common are the Gravure Research Institute (GRI) for paper and the Diamond National for paperboard. These testers print a grid of halftone dots, and the smoothness is evaluated by the number of missing dots. If the paper is smooth, fewer dots will be missing. These tests are limited by the fact that they test only a small amount of the total surface of the web, and rely on human judgment of quality, although the use of image analysis to count the missing dots improves the quality of measurement.

Ink Receptivity and Holdout

In interaction of the paper with ink is equally as difficult to measure as smoothness. The closest to a standard test is the use of the *K&N ink test*. This ink is a gray-colored oil that is spread on the surface of the web, allowed to soak in for 2 min and then rubbed off. Many prefer the red Croda wipe ink, which is used the same. The Croda seems to show differences in the coating best and the K&N shows basestock differences best. The ink penetrates the paper and makes it darker. The ink receptivity is then measured by how much darker the inked area is than the uninked sheet. Again, the use of a small area and of ink that is different from the actual printing inks limits the reliability of these methods. Many proof-printing schemes have been devised and are being used. Most methods involve printing the paper and observing the gloss of the ink after a certain time interval. Each method can potentially be of benefit for a certain grade or process for which it is specifcally designed.

Strength

If the paper is to be printed in a web-fed press, the strength of the web is of concern to maintain production and avoid web breaks. *Tensile strength* of the web is seen as a basic strength evaluator, but all too often web breaks are not caused by low strength. Most web breaks come from: impurities in the web, which cause weak spots; damage to the web during processing or shipping; tears or cracks in the web; or wrinkles, which require the printer to use too much tension on the web during printing. Web breaks can be avoided only if the web has sufficient strength to overcome these incidental factors or if all of these incidentals are eliminated.

Offset printing subjects the surface of the web to a vertical lifting force known as a picking force. *Pick strength* can be evaluated by several methods, but the most common is the *IGT pick tester*. This apparatus prints the sample with an ink or oil of known viscosity, determining at what picking force level the web will fail. The accuracy of this test is limited by the small sample measured and the differences between the test ink and the inks and dynamics of full-size presses. However, the tests can be empirically correlated with actual performance and made meaningful.

REFERENCES

The following texts are recommended for further information on printing methods, the relationship between paper properties and print quality or print evaluation methods:

Casey, J. P. Ed., *Pulp & Paper*, Vol. 4, New York, John Wiley & Sons, 1983.

Glassman, A., *Printing Fundamentals*, Atlanta, TAPPI, 1985.

Kocurek, M. J. Ed, *Pulp & Paper Manufacture*, Third Ed., Vol. 8 The Joint Textbook Committee of the Paper Industry, Atlanta, TAPPI, 1990.

9 Corrugating Operations and Raw Materials

From the standpoint of the paper and paperboard industry, corrugating board is the largest single grade of paper produced, and remains a growing grade. The universal use of corrugated containers for shipping manufactured goods makes this grade one of the largest forms of packaging from the standpoint of the packaging industry also. Manufacture of the corrugated board is such that there are a large number of converting plants located throughout the country, close to the users of the containers. The paper for these corrugating plants is made at papermills located in the traditional regions near the forests and water.

The corrugating operation uses two types of product from the paper industry: *corrugating medium* and *linerboard.* The medium is the fluted or corrugated center layer of the board. Standard 26 $lb/1,000ft^2$ medium is made 0.009 in. thick, which causes it to be called 9 point board. The linerboard is the material used on the top and bottom of the sandwich. Corrugated board is made in a variety of forms. The most common, that which has two liners and the fluted medium, is known in the industry as *double face.* If there is only one liner, the product is known as *single face.* Single face is frequently used as a liner inside the container to help cushion bottles and other materials. If single face is glued to a double face so that liners are exposed on both faces, the product is called *double wall.* The double wall allows the use of two different-size flutes to give the box added quality. As shown in Figure 9.1, the flutes are made in four different sizes. The largest is A flute, which makes the board the thickest and gives the container the greatest stacking strength. The smaller flutes have greater puncture resistance. By combining both in a double-wall board, we can combine both properties in the container.

It is possible to glue more than two layers together, making more than a double-wall board. The resulting multiwall or multiple-layered material is extremely stiff for its weight and is used as reinforcing at corners to replace wood or can even be used to build skids or other structural pieces. These products are specialized and do not constitute a major portion of the corrugating business.

THE CORRUGATOR

The Single Facer

The first part of the corrugating machine is known as the *single facer*, since it is here that the medium is fluted and attached to the top liner, making single-face board.

Flute	Number of flutes[1] per		Flute height[2]	
	Lineal foot	Lineal meter	inch	mm
A	36	118	3/16	4.76
B	51	168	3/32	2.38
C	39-42	128-138	9/64	3.57
E	96	316	3/64	1.19

[1]Approximate
[2]Exclusive of liners

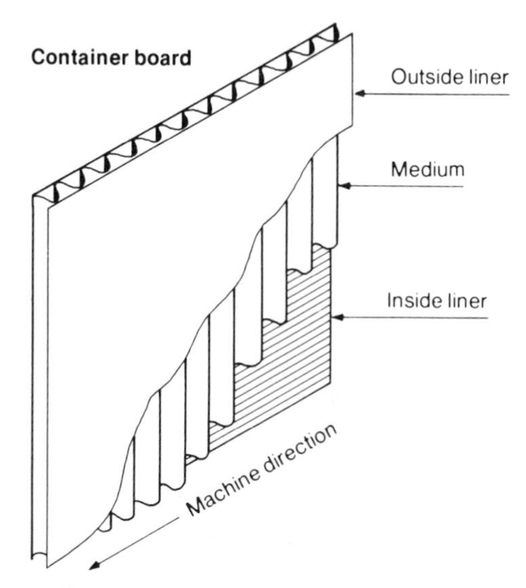

Figure 9.1. Corrugated board flute sizes (Source: Lowe 1975, p.100)

The operation can be divided into four basic sub-operations: (1) the unwinding and conditioning of the two webs, (2) the actual corrugating of the medium, (3) combining and gluing the two webs and finally (4) holding the single face in the bridge to allow the glue to set or dry.

The drawing in Figure 9.2 shows the medium being supplied from an unwind, which is under the part of the machine known as the *bridge*. The top liner is unwound from a position at the outside of the machine. The unwind stands are generally turret unwinds, which allow the use of flying splices so that the rolls can be changed without stopping the machine. Both webs may be passed around heated drums, to heat them before combining. Steam may also be used, especially on the medium to increase its flexibility. If the two webs are at greatly different temperatures or moisture contents before combining, the resultant structure will likely curl. Primary responsibility for the moisture content lies with the manufacturer of the webs, but small adjustments can be made on the machine.

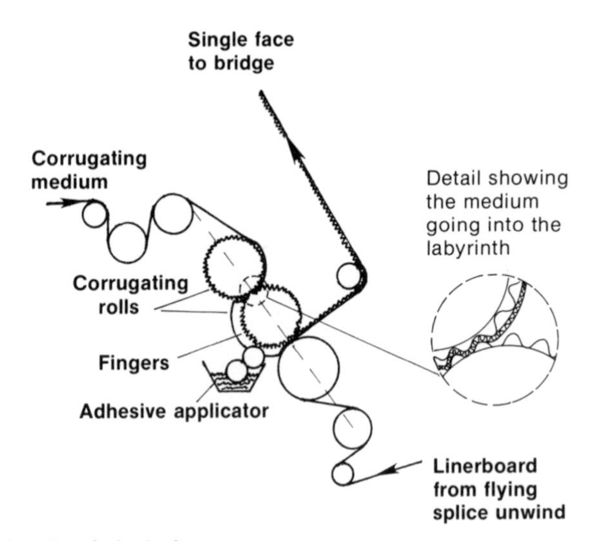

Figure 9.2. Details of single facer

The web to be corrugated is passed between two rolls with corrugated surfaces or intermeshing teeth all across their surface, which force the web into the fluted shape. The web must be made to hold this shape after it passes from these two rolls and the contorted space between them, called the *labyrinth*. As it enters the labyrinth, the web is sprayed with steam to heat and moisten it to make it more pliable. The corrugating rolls are heated to dry the web and make it hold its shape. The inclusion of wax and starch in the steam shower is purported to improve release from the rolls and help to stiffen the medium.

As the corrugated web passes from the labyrinth, it is held against one of the two rolls by a series of thin steel plates called *fingers*. Between these fingers is a set of rollers used to apply adhesive to the tops of the flutes. Before the fluted medium and the adhesive leave the corrugating roll, the top liner is pressed against the glued surfaces. The adhesive is a starch solution of the proper solids and composition to become sufficiently tacky in this short time to hold the two webs together as they pass up to the bridge. This adhesive is generally a mixture of starch granules and pasted or cooked starch in which the granules have been broken down. Higher levels of water resistance require a latex type adhesive and a top liner to which wet-strength resins have been added. As was noted earlier, the board is taken up to the bridge, where it is stored to give the adhesives time to dry and form a permanent bond.

The Double Backer

If double-face board is to be made, the single face produced by the operations just described must be glued to another liner. The back or bottom liner is unwound from a similar flying splice type unwind also found under the bridge. The bottom liner may also be sprayed with steam or heated to increase the similarity between it and the rest of the assembly, to reduce the tendency to curl. As can be seen in Figure 9.3, the single face is brought from the bridge and adhesive applied to the tops of the exposed flutes. The pressure used in this nip must be controlled to ensure that the flutes are not crushed, but that all flutes receive enough adhesive to bond the board together. The top of the single face may also be sprayed with steam to help equalize the board moisture content and prevent curling.

Shortly after the adhesive is applied to the single face, the back liner is brought into contact with it and the assembly is fed between two continuous belts into the steam chest. The adhesive used for this operation is similar to the first, but need not be capable of developing the quick or green tack, since these layers are held together under pressure and heat to dry and cure both glue lines. The pressure is sufficient to hold the layers together, but must not be so great as to crush the flutes.

To make double-wall board, another single facer, bridge and double backer gluing station would have to be installed on the machine. In fact, most corrugating machines are built with two sets of corrugating rolls and provision for making double-wall board. The addition of the second set of corrugating rolls has value other than just the manufacture of double wall. The two sets of rolls are generally in two different flute sizes, perhaps an A flute and a B flute. Both flutes combined in one double wall give it puncture and edgewise-compression resistance. The use of one size of flute at a time allows the manufacture of either of two different types of double face.

Conversion to Finished Blanks

After the board has been glued and cured into a continuous web of product coming from the drying or steam table, it must be cut into individual sheets, or *blanks*. The corrugated board cannot be wound onto a drum for further processing or it would become permanently curled. As the web comes from the steam chests, it

Figure 9.3. Details of double backer

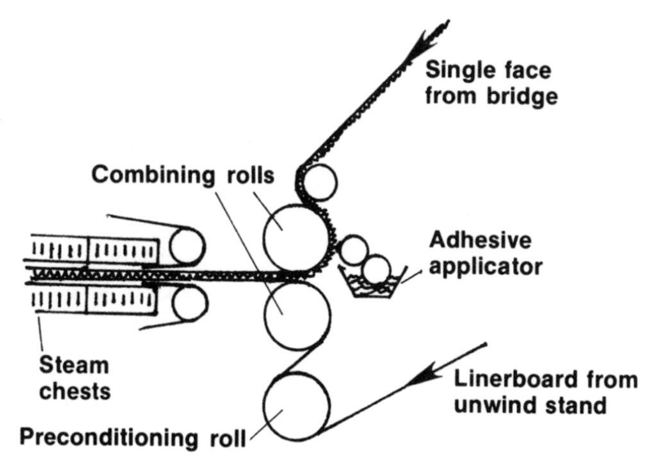

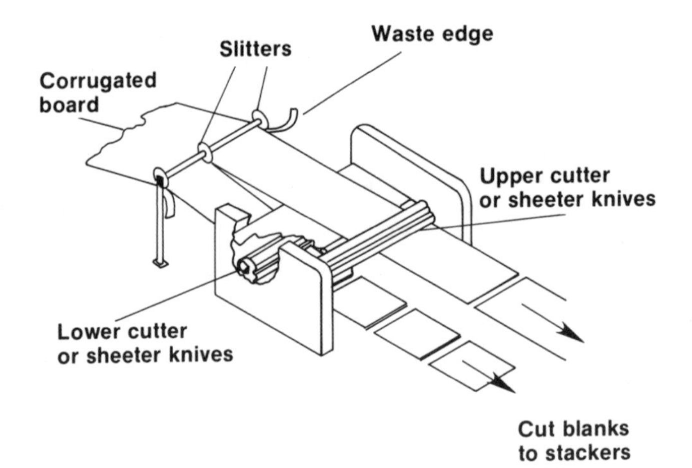

Figure 9.4. Trimmer-cutter details

first passes through a slitting and scoring operation as shown in Figure 9.4. This device has a series of sharp discs positioned above the web that can be lowered into the web to slit it into narrower webs. The slitter knives are movable to allow a variety of web widths to be cut. This device can also be used to score the board in the machine direction by replacing the sharp discs with non-cutting dull discs, which are pressed into the surface of the web. The webs are then cut into sheets by one of two sheeter knives. The machine is normally fitted with two sets of sheeter knives so that two different sheet lengths can be produced. The machine therefore also needs a layboy stacker to accept the cut blanks and stack them in piles of different-size sheets.

CONTAINERMAKING

The actual making of the corrugated board is an important component of the corrugating plant's work, but is by no means the total operation. The plant sells containers, not just corrugated board. The major portion of the plant is occupied by equipment to convert the blanks coming from the corrugator into containers, or some form of saleable package. Profitability of the operation relies on efficient use of the corrugator capacity and also on conversion of the board into printed, folded cartons. As the board comes from the corrugator, it is cut into rectangular blank sheets, which may be scored in the direction perpendicular to the corrugations. Since the maximum stacking strength is in the direction of the corrugations, that will be the vertical direction of the finished container. To be converted to a container, the sheets must be printed, cut and creased to form the folds and tabs that will be glued together to form the container. A large variety of machines are used and it is impossible to describe them all here. A few of the more popular devices will be discussed as examples of the common operations performed on corrugated board.

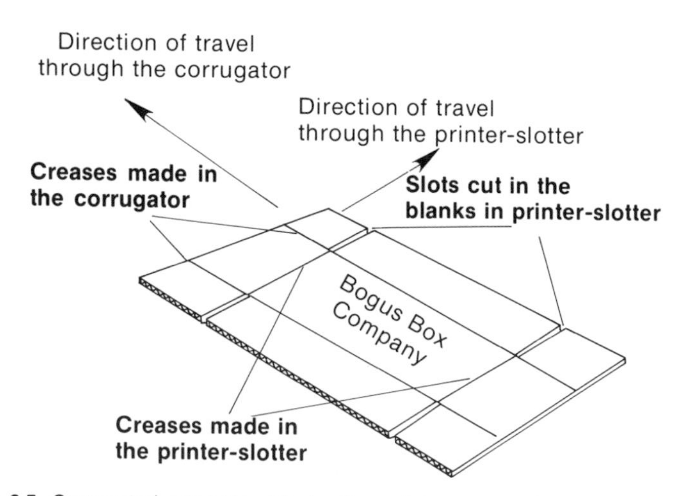

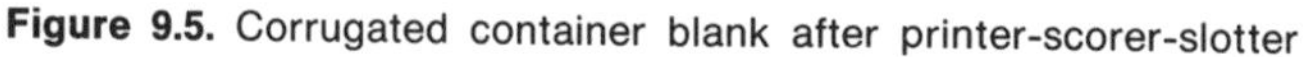
Figure 9.5. Corrugated container blank after printer-scorer-slotter

The Printer-Slotter

A printer-slotter is used to print the blanks and cut the tabs, but will not fold and glue the container. The printing is normally done with a flexible letterpress type plate, using a fairly oily ink of the letterpress type. The sheets are passed through the press in a direction perpendicular to the direction of travel in the corrugator, producing an unfolded container like the one shown in Figure 9.5. If the design of the container is proper, this press can produce a container ready to be folded and sealed. The sheet will have been scored in one direction in the corrugator; by going through this machine in another direction, as shown in Figure 9.6, it can be scored using a pair of rollers on either side of the blank, as shown in Figure 9.7. Small slots can also be cut to create the tabs needed to seal the container. The purchaser of the containers may want them shipped in this condition, intending to fold and seal them in his own plant, or the corrugating plant may need to do the job. The blanks can be folded manually or automatically, and the tabs attached by gluing, taping or stitching with staples. If the boxes are to be glued, the job is more likely to be done on a machine that can print, score, fold and glue all in one pass. Taping and stitching are generally done separately, as they are slower operations and not as readily adaptable to in-line operations.

Figure 9.6. Schematic of printer-scorer-slotter showing the sequence of operations performed on corrugated blanks

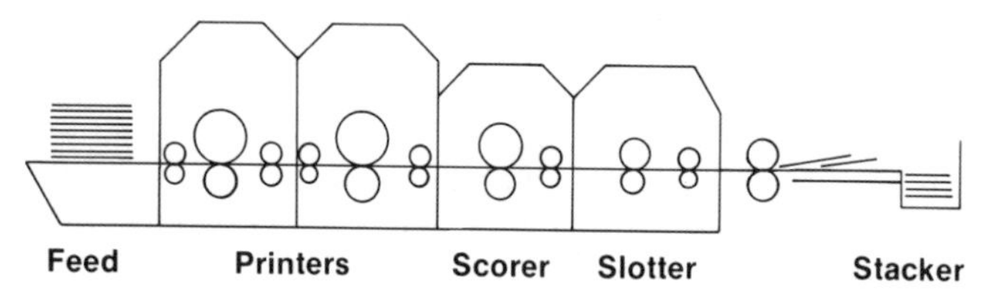

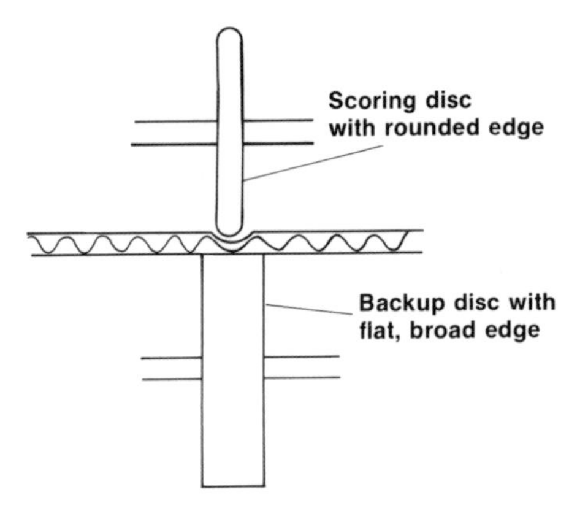

Figure 9.7. Scorer design

Diecutters

If the container design is more complex—requiring slots in both directions, at many different places and at angles to the edges of the sheets or requiring punch-out windows—the container must be cut differently. A diecutter uses a curved diecutting plate, with sharp knives and dull scoring blades extending from its surface. The diecutter may also have attached or in-line printing stations to allow blank sheets to be printed and diecut in one pass. The blanks still require folding and sealing in a separate operation. Depending on the complexity of the folding operation, the folding may be done in a machine or manually, as in the case of store display stands.

Printer-Slotter-Folder

Printer-slotter-folders have become quite complex and sophisticated. Sheets from the corrugator are loaded into one end and finished containers are delivered to skids and perhaps even wrapped and sealed for shipment on the other end. The streamlined, efficient operation of this machine makes it the preferred boxmaking machine, and the most profitable to operate. Designs are fairly flexible, but still limited to scoring and tab-cutting in the direction of the flutes, or perpendicular to the score made on the corrugator.

Printing is accomplished with flexo presses using water-based inks. The press may print up to four colors, although most are equipped for only two or three. The brown color of the container is not well suited to four-color printing, but by properly selecting the colors, some interesting graphics can be created. The flexo plates are stapled to a wooden printing plate cylinder, and can be changed rapidly when jobs are changed. The ink is pumped up to the printing unit from a 5-gal pail set on the floor next to the machine and can also be changed rapidly. The flexo printers use the anilox roll modification and are capable of halftones of fairly good quality. The use of flexo allows the plate to be pressed against the board, conforming to its rough surface, with little danger of crushing the flutes.

The slotter-scorer section has recently gone through a major advance in automated setup to allow faster changeover from one order to another. The scorers are dull discs attached to a shaft above the board, as shown in Figure 9.7, and the slotters are U-shaped dies, which also are attached to a shaft above the board, as shown in Figure 9.8. The slotters and scorers need to be positioned at the right places from side to side on the machine, and the slotters also have to be placed at the proper location with respect to the horizontal travel of the blank through the machine. The positioning is a slow, critical and painstaking task to accomplish by hand. The operator must individually measure and set all slotters and scorers in what is believed to be the right place. A few sample blanks are then sent through, measured and the settings corrected if needed. The newer machines are equipped with servo-motors that can position the scorers and cutters. The position of each device can be dialed in or fed to the controller by card, the devices automatically positioned and the sample blanks run. It may still be necessary to make minor adjustments, but they also can be dialed in and made by the servo-motors. Since most plants operate on small orders requiring frequent changes and must deliver in a hurry, these advances are most helpful.

The printed, scored and slotted blank is now ready to be folded and glued. The adhesive is normally a latex type, which can be applied by rollers or, more frequently, sprayed onto the tabs as they pass through the machine and under the spray heads. The positioning and timing of the spray are adjustable and can be automated. The tabs are folded and pressed into position by metal tabs and rollers. The folded container is generally fed between belts, where it is held closed briefly, allowing the adhesive to set. The container is also fed to a stacker, which keeps it closed until it can be loaded on the skid. More advanced machines collect, count and position the stacks on the skid and prepare the skid for banding, wrapping and shipping.

Figure 9.8. Slotter design

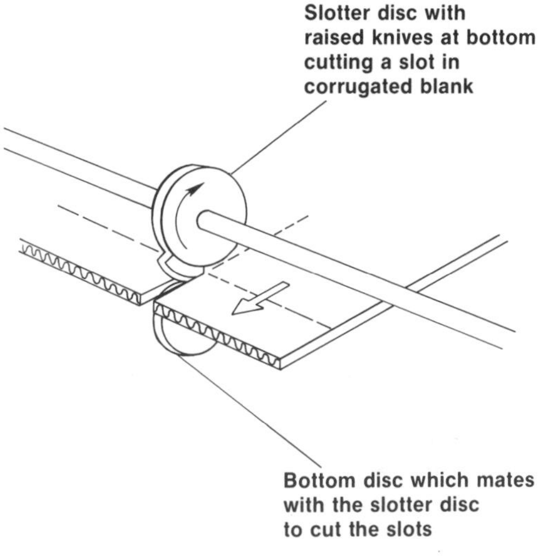

CONTAINER AND CONTAINER PLANT DESIGN

The corrugated container plant is generally a small operation, located near the user of the containers. The bulkiness of the product dictates that the cost of shipping will be high, and the storage of large numbers of containers by the user is also prohibitive. The containers are therefore ordered locally and in small quantities. If there is more than one corrugator in the area, competition soon becomes quite sharp. Most plants have design shops to help modify designs so that they can be made with less board—for greater efficiency in their own plant and, therefore, a lower sale price—or to offer special graphics or processing advantages for the user. The corrugated container comes in millions of forms, each designed to have some special advantage over the others. The design of the plant to facilitate flow of materials and operations is also extremely important in cutting cost and improving profitability.

PAPERMAKING FOR CORRUGATING

Corrugating Medium

The predominant raw material used in the manufacture of corrugating medium is semichemical hardwood. Hardwood pulp is used rather than softwoods because the hardwoods are less costly and contribute to the particular type of strength needed in corrugating medium. The long fibers of the softwoods are generally considered to make stronger paper, but in this application the long fibers are not needed. The pulp for medium is what would be called a *raw cook*, meaning that the lignin is not totally removed. Furthermore, the pulp is not washed thoroughly, or in some cases, is hardly washed at all. By not washing or using a strong cook, the lignin and other hemicelluloses are left with the fibers and will be formed into the web of paper. When this web goes through the corrugator, these chemicals help form the rigid fluted shape needed. The short hardwood fibers are less flexible than softwoods and therefore contribute to the stiffness of the fluted structure.

The processing starts with the hardwood chips, sometimes containing some bark when whole-tree chipping is used; other times the hardwood chips are obtained as a waste by-product of lumber or furniture manufacture. In either case, the chips are cooked with some form of high-yield cooking process. Neutral sulfite semichemical is perhaps the most popular, but modified kraft and soda cooking are also used. Continuous cookers are common, but some of the older mills still have batch type digesters in use. The chips are mixed with the liquor, sent through the digester, broken down by disc refiners, squeezed or washed to extract most of the cooking liquor and sent on to the papermaking operations. The liquor removed in the extractors can be fortified with fresh chemicals and reused. In operations where both the medium and linerboard are being produced in the same plant, the waste liquors from both can be mixed, sent to a recovery boiler and fresh liquor generated.

The stock coming from the pulping operation will need to be cleaned and screened to remove uncooked knots, nails and other foreign materials, but will not require much refining or addition of other materials. Hardwoods do not respond well to refining in the first place, and refining to improve bonding in the web is not as essential since the lignin and other chemicals still in the fibers will help bond the web together. Fillers and sizing agents are also not needed in this type of paper unless wet-strength agents are added to make special water-resistant containers.

Secondary or recycled fibers are also used in the manufacture of medium, requiring that the mill have some form of pulper and storage facility in the stock prep area. The common wastepaper used in medium is old corrugated containers or cuttings from corrugating plants. However, the level used cannot be much above 40% of the weight of the paper or the stiffness will suffer. Corrugating medium made from 100% wastepaper, principally from old corrugated containers, is called *bogus medium.* For an equivalent use, it is made in somewhat higher basis weights to compensate for its lower strength.

The medium is commonly made on fourdrinier type paper machines. The machines do not need size presses, but will be equipped with calender stacks. The basis weight of medium and linerboard is expressed in pounds per thousand square feet, and for medium will range from 26 to 33 lb/1,000 ft^2. The thickness of the medium is perhaps most important, since it must be able to fit in the space between the flutes on the surfaces of the corrugating rolls. The medium is also called 9 point board since its thickness is 0.009 in. In addition to caliper and basis weight, the other important properties are: moisture content, mullen strength, tensile strength and machine-direction stiffness. The flutes in the board can be considered to be like small truss girders in bridge construction. The crushing load exerted on the board attempts to bring the two liners of the board together by collapsing these struts or flutes. Since the flutes are made with the machine direction running up and down in them, the machine direction stiffness is most important in preventing crushing of the board. This property is also tested by forming a small sample of corrugating medium in a machine known as a *Concora tester.* This machine takes a thin strip of medium, flutes it and attaches it to a piece of tape on one side. This assembly can be placed in a tester, which will crush the flutes and report the force required. The cross-machine direction, the long dimension of the flutes, requires stiffness for the overall stacking strength of the container. The construction of the container is such that the flutes will run up and down the vertical sides of the container, since the flutes make small tubes in the sides and tubes are strong support members. Stiffness of the paper can also be tested by forming a single coil or tube of the paper about 1 in. high and measuring the force needed to crush this tube. This test, called the *ring crush*, can be either run in the machine or cross-machine direction.

Linerboard

The linerboard is made from predominantly softwood fibers, but may contain up to 20% hardwoods or secondary fiber. Again the secondary fiber will be cuttings from the corrugating plant or other waste corrugated board. The softwoods are needed here to give the linerboard the necessary strength. The kraft cooking process is the most common, but high-yield processes have been proposed and may be used. The pulp is cleaned and washed only enough to reclaim the waste chemicals, but not to improve the color or brightness of the board.

The pulp for the liner will need to be refined to help improve the strength of the board. The stock prep area will need to have some form of pulper to defiber the secondary fiber, if used. Mineral fillers are not used since they can be harmful to the strength of the board; however, as the cost of fiber sources increases, there is a greater possibility of the use of clay in the board to add to the weight, reduce the cost and perhaps improve printing characteristics as long as the strength is not hurt too much by this addition. Chemicals may be added if board with water resistance or wet strength is needed. Where melamine formaldehyde had been used, the need to remove formaldehyde from the environment has led to its replacement by other chemicals, such as amino polyamide-epichlorohydrin resin.

Linerboard is commonly made on a flat-wire fourdrinier in a basis weight range of from 26 to 90 lb/1,000ft^2. The lighter board can be made easily on a normal fourdrinier, but the heavier weights are usually made in two layers by the addition of a secondary headbox. As shown in Figure 9.9, the web is delivered to the wire first from a standard headbox at the breast roll. When the web delivered by this first box is quite well formed, near or at the dry line, another headbox mounted above the wire delivers more stock on top of the already formed web. This added stock may drain through the already formed web, or a second wire may be brought down on top of this stock so that water can be removed from the top of the combined webs as well as through the bottom. The use of the secondary headbox allows a thicker web to be formed without sacrificing machine speed or quality of formation of the web. Furthermore, the secondary headbox allows the use of a different furnish in the top layer. The top layer can contain more long fiber for scuff resistance, hardwood for printability, or bleached white fibers (called mottled white board) for looks and printability.

As with the medium, important properties of the linerboard are: basis weight, moisture content, caliper, mullen and tensile strength and cross-machine stiffness. As described earlier, the tubes formed by the flutes run in the cross-machine direction of the linerboard as well as the medium. Therefore, the stacking strength of the container comes from the cross-machine stiffness of the linerboard and the medium.

Figure 9.9. Fourdrinier with secondary headbox

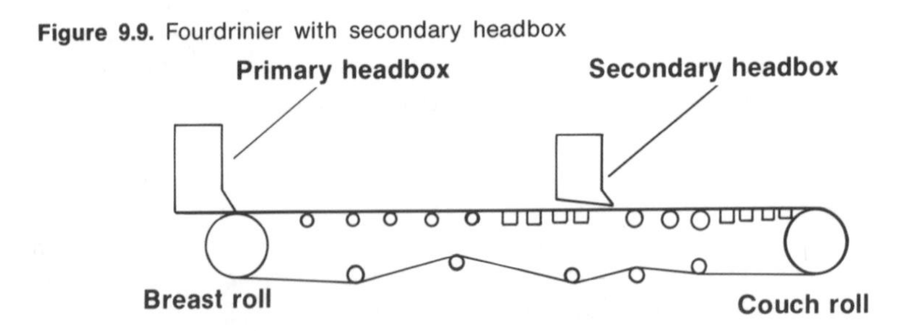

View of fourdrinier showing secondary headbox.

New Trends and Special Considerations

The major trends affecting paper manufacturing are in the area of potential fiber shortages and the growing need to use secondary fiber in the manufacture of the medium and linerboard. The corrugating plant is under tighter economic pressure to maximize efficiency in all phases of its operation. First of all, this is seen in the development of controls and splicers that allow continuous operation of the corrugator itself. The development of automatic setting controls on the slitter, scorer and cutters as well as improvements in the layboy to improve and speed delivery of cut blanks from the machine are also important to reducing lost time when changing orders. Conversion of the corrugated blanks into finished containers is also becoming faster and more automated by the use of more in-line operations and automatic or remotely set slitters and scorers.

Improved printing characteristics of the linerboard will continue to be important as the current trend toward self-service shopping centers continues. The use of mottled white or solid white linerboard has increased to satisfy this demand, and is expected to continue. The white fibers are supplied by secondary fiber sources in most cases. Another recent trend is to coat the linerboard. The coatings are applied directly to the unbleached linerboard and will contain a high ratio of titanium dioxide or other pigments to promote the ability to hide the unbleached stock and promote brightness. High-quality web offset printing may be done on the linerboard prior to corrugating, or on the boxes in the normal manner. The flexographic printing process remains the most common for high-quality printing on the finished boxes and is expected to continue to dominate.

REFERENCES

Casey, J. P. Ed., *Pulp & Paper*, Vol. 4, New York, John Wiley & Sons, 1983.

Kocurek, M. J. Ed, *Pulp & Paper Manufacture*, Third Ed., Vol. 8 The Joint Textbook Committee of the Paper Industry, Atlanta, TAPPI, 1990.

PAPER GRADES AND MANUFACTURING OPERATIONS

Raw Materials and Pulping Methods

Most packaging grades of paper are made from softwoods, to take advantage of their greater strength potential. The bleached grades used for labels and overwrappers, which are intended more for their decorative capabilities than for their strength, may include some hardwoods, mechanical pulp or secondary fiber. Low permanence (tendency to yellow) severely limits the potential for mechanical pulps, but all of these raw materials are noted for their ability to increase bulk of the web and improve printability. For the pulping process, kraft is the most commonly used, again because it contributes to greater strength in the paper. The use of semichemical pulping processes is possible in applications such as cans and drums, where the paper will be glued into a structure in which some of the strength and stiffness comes from the adhesive and the strength of the paper is not as important. The pulps are used both bleached and unbleached; the unbleached kraft is easy to identify by its brown color. As indicated in Table 10.2, the bleached grades are used where printability, decorativeness or whiteness are needed. Unbleached and recycled grades can also be made printable and decorative by pigmented coating.

Secondary fibers are used in almost all grades, at least as broke internally circulated within the mill. Bag applications, where tearing strength is of greatest importance, are least likely to use secondary fibers. Printing overwraps in the bleached grades and can, drum and core applications in the unbleached grades are most likely to use secondary fibers. Many of the unbleached can or core grades make extensive use of secondary fiber, especially corrugated waste, which still has good strength and gives the characteristic brown color.

Stock Preparation Operations

Stock preparation is relatively simple. Most grades are refined to develop maximum strength. Mineral fillers are used primarily in the printable bleached grades, and almost all grades are sized internally, even those listed in Table 10.2 as having no treatment. The treatment referred to there is web modification or other converting operations.

Glassine and Hydrated Paper

Glassine and hydrated paper are the notable exceptions to the general statement that nothing special is done in the stock prep area. By refining the fibers beyond the minimum necessary to develop normal strength levels, the web becomes more dense and develops some oil and grease resistance. The normal web of paper is rather porous, having about 50% of its volume occupied by air spaces between the fibers. Refining breaks the fibers down, exposing more fibrils on their surfaces and making them more flexible. As these fibers are formed into a web, the more flexible fibers will flatten and form a more dense web. The increased fibrillation will also help fill in the holes in the web, creating more resistance in the web to

Table 10.2. Paper and Paperboard for Packages

	Function or use	Treatment
Paper grade		
Unbleached kraft softwoods	Strong, flexible containers Bags, drums, cans, boxes	None
	Barrier and strength	Coated or laminated
	Stretch and strength: cushion	Creped or extensible
Bleached kraft softwood	Fancy bags, envelopes, labels	None
	Overwrap and labels	Clay coated
	Greaseproof; wrapper, box	Highly beaten pulp (hydrated)
Glassine	Grease and oil resistance Wrapper, liner, deli	Hydrated and supercalendered
Parchment	Grease, oil and water resistance Wrapper, liner, release, deli	Acid-treated paper
Any of the above	Barrier and strength	Coated or laminated
Paperboard grade		
Corrugated medium and linerboard	Corrugated containerboard	Corrugating and assembly
Solid bleached sulfate (SBS or fourdrinier board)	Stiff, strong, white Frozen food boxes, trays	None
	Barrier and strength Milk or ice cream	Plastic coated or laminated (extruded)
Solid unbleached sulfate (SUS)	Stiffness and tear strength Heavy-duty boxes, beverage carriers	None or clay or plastic coated and/or laminated
Chipboard and vat-lined chip	Stiff, strong Setup boxes or folding box	None: perhaps used with liners
Clay-coated and bending chipboard and lined	Stiff, strong, printable Folding boxes	None or plastic coated and laminated
Multifourdrinier board (MFB)	Stiff, strong, white, printable Folding boxes, trays, etc.	Bleached kraft/TMP*/ bleached kraft

*TMP=Thermomechanical pulp

penetration by oil or grease. Hydrated paper is made by excessive refining of the pulp. Glassine is made by sending hydrated paper through supercalenders to further compact the web and increase its barrier characteristics. As previously explained, the supercalender is a stack of rollers in which steel rollers alternate with rollers filled with cotton or denim pressed so hard that no resilience is noticeable in the rollers. The filled rolls are actually softer than the steel and give slightly as the web is passed through, causing a polishing action on the surface of the web. These grades are sometimes waxed or plastic coated to further improve their barrier characteristics.

Web Forming Operations

Almost all paper for packaging is made on the normal flat wire fourdrinier paper machine. Some mills have twin-wire machines, but they are the exception. The other exception to this general statement is in the manufacture of paper for tubes, drums or cores. The paper for these grades is usually categorized as paperboard and is made on either cylinder board machines or multiwire, fourdrinier type machines.

Web Modification Operations

Physical Modifications

Web modifications such as sizing and calendering are commonly used on the white grades and may be used on the brown grades as well. The major modification operations used for these grades go beyond these normal paper machine operations, however. Physical modification is used in creped grades, formed by drying and creping on a yankee dryer much the same as tissue grades. Extensible paper is treated to a special process in which the moist web is pressed against a rubber surface that has been stretched. As the rubber is allowed to shrink back, the web is compressed on its surface, creating a condition called *microcreping*. This treatment, although slow and expensive to perform, greatly increases the extensibility of the paper with little loss of strength. Extensible grades are used primarily for industrial-size bags to ship heavier quantities of bulk dry powders, such as cement. Extensible or creped unbleached kraft is used as tape backing, where strength and stretch are also needed.

Pigmented Coating

Most of the printing grades of both bleached and unbleached papers are pigment coated to improve printing characteristics. The coating of the bleached grades is rather straightforward, using most of the common pigments discussed in Chapter 7. Adhesive selection must consider the final application of the product; the adhesive may need to be waterproof or compatible with inks or glues used in subsequent laminations. When coating the low-brightness unbleached grades, there is a greater likelihood of multiple applications and the coatings may have a relatively high titanium dioxide content to improve their opacity and help hide the brown substrate. Multiple coating allows heavier coat weights to be applied while still using blade coaters, which are faster and more economical to operate than air knife coaters. However, the contour-coating capability of the air knife coater is better able to hide the base than the blade, which will fill in the valleys in the surface and therefore give a coating layer of varying thickness.

Because the paper is to be used in packaging applications, there is a great likelihood that the paper will only be coated on one side. This factor can create problems due to the greater tendency of coated-one-side papers to curl. The coating on one side creates an instability in the web, both at normal room moistures and when the web is subjected to water or even high-relative-humidity conditions. The coated web will normally tend to curl toward the coated side due to shrinking of the coating on that surface during drying. As the web picks up moisture from the air, the uncoated side is most likely to take up moisture and swell faster, again causing the web to curl toward the coated side. There is little cure for this condition, except to coat the back side of the web, which may be undesirable for other reasons. The paper coater will attempt to apply and dry the coating in such a way as to minimize this curling tendency, but it cannot be eliminated entirely.

Functional Coating and Laminating

As has been noted earlier, paper has little barrier capability other than to light, and if barrier characteristics are needed, the web must be coated or combined with something that will give the composite the barrier needed. Coating methods and materials are quite varied, ranging from a light wax coating applied in the paper machine to conversion applications of wax, varnish or hotmelts. Hydrated paper and glassine used for wrappers are most likely to be wax treated only since they already have some resistance to grease and oil. Box liners, pouches to be used in boxes, and other applications for dry foods may also get by with a light wax or plastic coating.

Plastic coatings may be applied as lacquers or varnishes, using a volatile organic solvent, or they may be applied as an aqueous emulsion. Polyvinylidene chloride (PVDC), a polymer readily available as an emulsion, is widely used for grease and oil resistance and waterproofness.

In applications where the barrier characteristics must be great, such as snack food bags, freeze-dried foods, tobacco pouches, and so forth, the barrier obtainable from coating may not be sufficient to do the job. In these cases the paper is commonly laminated to a plastic film or foil. The laminating can most easily be done as a combied extrusion-laminating process where the plastic film is extruded and the paper web (and perhaps also the foil) are combined while the extrudate is still hot and can function as the adhesive for the lamination. The film and foil contribute the barrier characteristics and the paper contributes the strength needed to contain and physically protect the product. Many of these packages contain no paper at all, when the films or the foil can supply the needed stiffness or strength or where the opacity of paper is not needed.

Special Properties and Tests for Packaging Papers

Among the basic paper properties discussed in Chapter 2 of this book, all of course are potentially important if they fall below acceptable minimum values. The most likely to be tested or listed as important for packaging papers are: basis weight, caliper, moisture content and directionality (machine to cross-machine differences). Optical tests of importance depend on the type of paper and the application. Printable grades need whiteness, brightness and opacity, while optical properties are not too critical in the unbleached grades. Strength tests of importance are primarily tensile, mullen, tear and elongation. All of these properties are covered by TAPPI and ASTM test methods. Stiffness and machinability are also important, but not as easily tested or specified. There are several machines available to test for stiffness, but the proof of the paper is in its ability to feed through the packaging machinery or function in the market. Machinability also is a combination of low surface abrasion or high slip, stiffness, freedom from curl, and so forth—all of which combine to facilitate conversion of the paper into a product.

In the packaging area, water sensitivity tends to be included in the area of converting or end use properties such as barrier to oil, grease, water, water vapor, odor and other gases. Most of these barrier properties are tested by placing the paper or other composite material in contact with the material to be resisted, and measuring the amount that penetrates into or through it. Vapor permeability can be measured by attaching a sheet of the packaging material over a dish of the liquid and measuring the amount that passes through the web over a given period of time. These tests can be performed hot or cold, and at different pressues for the vapors or gases to be tested. Although the paper may not be able to supply barrier characteristics, it can have a negative effect on them. If paper is to be coated to develop the barrier, the smoothness of the paper becomes very important. In order to obtain a barrier, a continuous film of the plastic material must be deposited on the paper surface. The rougher the paper, the thicker the coating must be to completely cover it. In many cases, extrusion lamination, which uses a continuous film of extruded plastic, can allow a good barrier to be developed with a relatively rough or porous web.

PAPERBOARD GRADES AND MANUFACTURING OPERATIONS

As indicated in Table 10.1, paperboard grades may be used in either folding cartons or setup boxes, indicating a rather simple grade structure. In fact, there are many grades of paperboard available. The most common, and easiest to describe, breakdown in grades is based on the type of pulp or fiber used to make the paperboard. Paperboard produced on cylinder board machines has commonly been made from secondary or recycled fibers and the fourdrinier machines have been used for virgin fiber. This division is actually quite basic and understandable when one considers the nature of the two different forming methods. On the fourdrinier, the entire thickness of the web is determined by the amount of stock coming from the headbox to the wire. With the cylinder machine the web is formed in layers, which are pressed together. The use of multiple layers on the cylinder machine allows the web to be made of several different types of fiber, while the fourdrinier did not have that capability until the recent development of special headboxes. A fourdrinier machine is not able to produce a multilayered web of recycled fibers as well as a cylinder machine can, and a cylinder machine cannot produce a web of all one fiber as well as the fourdrinier.

Advantages of the fourdrinier are higher operating speeds, which would be lost if recycled fiber were used, and freedom from delamination problems, which bother the multilayered paperboard. Advantages of the cylinder machine are in the ability to combine white outer liners with inexpensive gray filler layers, and the ability to produce thicker board than possible on the fourdrinier. Machines have recently been developed that are difficult to categorize as either cylinder or fourdrinier. Most of these machines are multiformers, but the formers are not clearly fourdrinier or cylinder. The machines are also mixed in that some are being used on recycled fiber and others on virgin fiber. We may have to redefine our grades as solid and multi-ply rather than defining them on the basis of the manufacturing operations. However, the user is more likely to want them to be classified on the type of fiber, brightness, strength or bending characteristics. In any event, the grades will be divided here on the basis of raw material, with the manufacturing operations as a secondary consideration.

Fourdrinier Board (Virgin Fiber Board)

Raw Materials and Pulping Operations

As in the case of the paper grades for packaging, kraft softwoods are the primary pulp selected because of their strength. The summary of grades in Table 10.2 shows that the grades are also divided into bleached and unbleached categories. The bleached grades are used in applications where the appearance of purity is felt to be important to the sale of the product. Beside the food applications shown, cosmetics and other high markup products that can afford the higher cost of the white board and desire its appearance and excellent printability use this board. The main advantages of the unbleached sulfate board are tearing strength, stiffness and water resistance or strength when wet. These properties make this an excellent material for returnable beverage containers or large-size soap boxes. The final grade listed—multifourdrinier board (MFB)—is a composite board using bleached kraft for top and bottom layers, and filling the middle of the assembly with bulky thermomechanical pulp (TMP). TMP, a mechanical pulp similar to groundwood, is made by steaming chips and passing them through a disc refiner. The resultant pulp is fairly bright, although slightly cream colored, without bleaching and can give the paperboard the appearance of solid bleached board. The stiffness of the MFB may not be any greater than solid bleached sulfate (SBS), but its cost is generally lower due to the TMP, and it is possible to make thicker board with the multifourdrinier than with a single fourdrinier.

Forming Methods

The basic fourdrinier used for most SBS and solid unbleached sulfate (SUS) is the flat forming section type, with the common inclusion of a secondary

Multifourdrinier machine at Continental Forest Industries' Augusta, Georgia, mill.

Beloit Bel Bond former at St. Regis mill, Battle Creek, Michigan.

headbox. The multifourdrinier grade listed (MFB) is based primarily on the production of one mill. A rough drawing of the forming section of the machine is shown in Figure 10.1. The bottom layer, or back liner, is depostied on the bottom wire first. The middle, or filler layers, are each formed on short fourdrinier wires above the main wire, then brought down and pressed against the back liner. The new webs are drawn from their individual forming wires by suction boxes under the main wire. The final wire and headbox apply the top liner.

Another design of multiwire former is shown in Figure 10.2. In this design, the stock from each successive headbox is deposited on the web and the water removed both down through the web and up through the secondary wire. This machine is usually used with reclaimed fibers to make solid unbleached grades or clay-coated lined chipboard. This machine and the multifourdrinier described previously are still exceptions. They have been highlighted to show some of the variability found in papermaking equipment; they also indicate the difficulty in categorization of grades by machine type, since both are technically fourdrinier board, yet both are also combination boxboard.

Figure 10.1. Multifourdrinier type combination boxboard machine

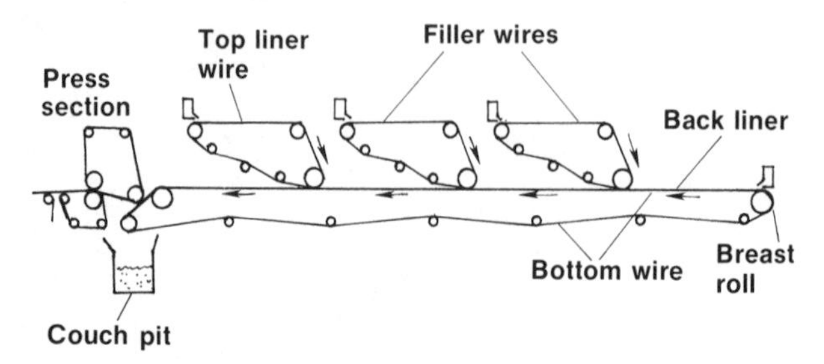

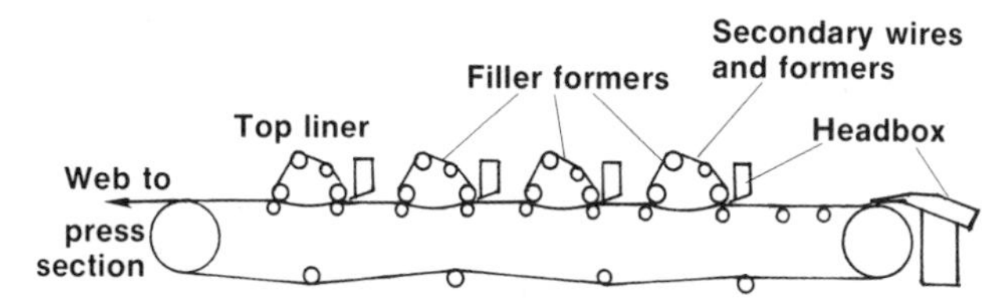

Figure 10.2. Beloit's Bel-Bond combination boxboard machine

Cylinder Board (Secondary or Recycled Fiber Boards)

Raw Materials

Cylinder board grades of paperboard are made from almost 100% recycled fiber. The common material is a combination of old newspapers, corrugated and carton cuttings which are pulped by mixing with hot water to reduce them to fibers. The pulped mixture, called *chipstock*, is cleaned to remove some of the foreign materials, but is not washed to remove the ink. The resultant paperboard is gray colored and flecked with black specks of ink. Paperboard made completely of chipstock is called *plain chipboard.* Plain chipboard is made in both bending and nonbending grades. The difference is in the stiffness of the paperboard and the ease with which it can be scored and bent. By using white wastepaper exclusively, white top liners can be added to the chip to make *white* or *vat-lined chip.* The chip is still used for the middle layers of the paperboard, where it is referred to as *chip filler.* The white top layer may need a second white layer under it (between it and the chip) to help hide the gray color of the chip filler. The last two cylinders of the cylinder machine would then apply an underliner and a top liner. The paperboard can be made stiffer and yet more easily folded by the addition of a kraft back liner. To do this, unbleached kraft from old corrugated containers or corrugated clippings are pulped and fed to the first vat on the cylinder machine. The longer unbleached kraft fibers make the board stronger and improve its folding characteristics.

Forming Methods

The cylinder formers described in Chapter 6 are the formers used to make the cylinder board grades. These formers are used as described in Chapter 6, Figure 6.3, to form and assemble a multi-ply web. The cylinder machine will have between five and eight cylinder formers, and all of the formers (with the possible exception of the top liner cylinders) will be of the same design. Some machines that were originally equipped with conventional cylinders (uniflow or counterflow) have been rebuilt and the top liner cylinders replaced to improve formation in these layers. Although newer machines may be built with a mixture of cylinder types, this is not a common practice. Most of the cylinder formers are designed to raise the web up after formation to present it to the bottom of the felt, which carries it on to the next former; therefore, most can be used together.

One notable exception is the Kobayashi Ultraformer, which forms the web on the cylinder and then delivers it to the top of the carrier felt, as shown in Figure

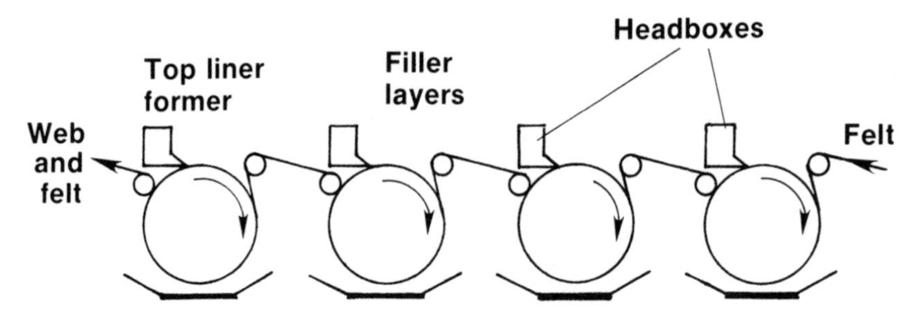

Figure 10.3. Kobayashi Ultra Former combination boxboard wet-end section

10.3. These formers are not easily combined with others in the cylinder machine wet end; when used, they will make up the entire wet end of the machine.

Web Modification Methods for all Paperboard Grades

Pigmented Coating

Any of the grades listed in Table 10.2 could potentially be coated with a pigment to improve printing properties. In most cases, though, only certain grades are coated. The industry convention is to refer to these as *clay-coated grades*, and they are so designated in Tables 10.1 and 10.2. The coating does contain clay as a major pigment, but it is not all clay. The discussion in Chapter 7 regarding general selection of coating pigments applies here in much the same manner. The major exception is in developing opacity. Since the coating is applied to a substrate that is already opaque or nearly so, the comments concerning the use of titanium dioxide must be modified. When the coating is applied to low-brightness paperboard, the titanium will make the coating more opaque—better able to hide the low-brightness substrate—and therefore will improve the final brightness of the coated paperboard, not its opacity.

The adhesive used in the coating of paperboard grades must be water resistant enough to be subjected to water during filling, as with frozen foods, or when placed in a sack with other moist or wet packages on the way home from the grocery store. Accordingly, protein and latex adhesives are common. Packaging operations also require glueability in the board, which is frequently favored by the use of protein and polyvinyl acetate type latex emulsions; however, since glueability is also dependent on the glue used, it is difficult to measure and predict.

Fourdrinier boards have most frequently been coated on the machine using blade coaters, while cylinder board has been coated by a variety of coating methods. Blade coaters are used with fourdrinier board because of a better compatibility between the speed of the machine and the capabilities of the blade coater. Some claim that cylinder board is too rough to be coated by a blade coater, but blade coaters have been used successfully on cylinder board. Cylinder board has been coated primarily with rod coaters and air knife coaters, however. Both have the proper speed compatibility to be installed on the cylinder board machine, and can do an excellent job of coating. The rod coater is used less because of a tendency to leave ridges in the coated surface, but is frequently employed as a precoater for the air knife. Recent developments in the use of low-angle, zero-angle and bent-blade coaters have raised the maximum coat weight obtainable with blade coaters to a range at which they can be considered for coating paperboard, which requires heavier coating than paper.

Functional Coating and Laminating

As indicated in Table 10.2, any of the paperboard grades can be given barrier characteristics by functional coating and laminating methods. As with the functional coating of paper for flexible packaging, the barrier comes from the plastic used either in the coating or in the extruded film, and the film or coating must be heavy enough to be pinhole free to give the desired barrier. Because of the rougher surface of paperboard, coating can be more expensive than extrusion laminating. The exceptions are in the use of inexpensive wax and wax-containing hot-melts for frozen food containers. Another potential advantage of coating over laminating is that the coating can be applied to the printed, scored and cut blanks. Thus, the clippings from the cutting are not contaminated with the coating and expensive coating is not wasted on them. Furthermore, the integrity of the barrier coating is not destroyed by the scoring operation when blanks are coated after scoring. Maintenance of the barrier in the folded portion of the carton is essential for the complete protection of the contents of the package. This characteristic is so important that milk cartons are sometimes coated after they have been formed into the final shape, just before filling.

Special Properties and Tests for Paperboard

Among the properties described in Chapter 2, basis weight, caliper, density and uniformity of all other properties, in both the machine and cross-machine directions, are probably the most important for paperboard. Basis weight is important to the user, who requires a certain surface area for his packaging operations. Caliper can influence the machineability, or performance, in carton-making machines. Caliper is also important in controlling density of the board, and can also have an influence on stiffness and scoring, as will be discussed later. The bulk is important in that bulk favors stiffness and also gives more package per pound of carton. Since the board is generally opaque, brightness, color and gloss are the important optical properties. Of the common strength tests listed in Chapter 2, tear and mullen are the most important. However, other strength properties such as *ply bonding* and *stiffness* are also important for paperboard. It is possible for the layers of the board to separate due to insufficient bonding between them. Ply bonding is helped by controlling the water contents of the webs as they are combined on the machine and by keeping the fiber contents of the layers similar in nature. The stresses incurred during printing and converting of the board to cartons can flex these layers and cause plybond failure. Failure can occur on printing presses, on gluing and folding lines or on filling machines.

Stiffness is a property needed in all grades of paperboard for packaging applications. Stiffness is generally measured with the Taber stiffness tester, which

er, and from hard white clippings for the top liner. As mentioned earlier, functional coatings are frequently applied after scoring to prevent damage to the coating during the scoring operation. The flexibility of the coating must also be compatible with the bending of the board during cartonmaking.

Another important property of carton boxboard is glueability. Adhesive is applied to one surface as the carton is folded into a flat, printed folded carton. To hold the carton together as it comes off the end of the cartonmaking machine, the glue must develop a quick or green tack, which must hold until the carton has been loaded into corrugated containers for shipping. When the carton is unfolded to be filled and sealed in the filling line, the adhesive must have cured and developed sufficient strength to hold together there and through the life of the box. To accomplish these two tasks, the adhesive must penetrate into the boxboard enough to increase in tack to supply the green tack. There are many different testers and testing schemes to attempt to predict the performance of the board. Most of these tests apply a measured amount of adhesive to one surface, press the two surfaces together under a controlled pressure and then pull them apart under controlled conditions. The pull-apart is generally timed to approximate the time the carton is in the folding line. Evaluation of failure is generally made in terms of the time elapsed before the tack has increased to the point at which the board, rather than the adhesive, will fail. This is referred to as *fiber tear*. There is usually a minimum time specified before fiber tear and a maximum time by which tear must occur. If the adhesive soaks in too fast, it may dry before the surfaces are brought together or it may remove too much liquid from the glue line to allow a good dry bond to form. Glueability is a complex property that is affected by the selection of the adhesive to be used as well as by the nature of the surface of the board. Smoothness, porosity and water resistance of the board surface are all of varying degrees of importance, depending on the dynamics of the gluing process. With increased use of hot melt glues, testing glueability becomes more complicated. Testers are being developed to apply hot melt glue and observe the same fiber properties. The porosity or absorbtiveness of the board is still important, and surface roughness is more important with hot melts.

Since most folding boxboard is used in some form of carton that will be printed, printability is very important to the overall performance of the board. The properties needed for good printability are dependent on the printing process to be used, but generally smoothness of the surface and proper ink receptivity are the key factors. Gravure and offset are the major processes used to print boxboard at the present time. Gravure demands good surface smoothness, but due to the drying of the ink immediately after printing, absorbency is not too great a problem. Offset printing is potentially a problem because of the high tack forces generated. The adhesive level needed in the coating to develop good glueability is normally high enough to ensure freedom from picking problems. The thinner ink film and longer drying times combine to give offset-printed board a tendency toward ink gloss problems.

General overall performance of the board in printing, cutting, scoring or any other conversion operation is of course also important. Flatness, dimentional stability and freedom from curling are all important in this area. These properties are also difficult to test and control. Paperboard must always be considered as a composite or layered structure simply because of its thickness. If any treatment is applied to one side and not the other, the stability is threatened. With combination boxboard, where the layers of the board are all slightly different, the problem is compounded. The basic response of fibers to water—possible expansion, curling or loss of stiffness of the paperboard—is probably the largest single source of converting difficulties.

11 Tissue and Related Grades

The tissue grades are seemingly quite simple and few, but the manufacturing methods are diverse. Toilet and facial tissues are similar, but not truly identical in either their manufacture or their processing and packaging. Napkins can also be made and processed in a variety of ways. Toweling includes both the kitchen towels sold in rolls and the white or brown towels folded to fit dispensers. Since the manufacturing operations are at least similar for all grades, the discussion will be organized on the basis of manufacturing processes. Special considerations for all grades that differ from the norm will be so indicated.

PULPING OPERATIONS

Softwood sulfite pulp has been the preferred raw material for most tissue and towels because of its softness. The sulfite process produces a softer, more flexible fiber, which gives paper the same characteristics. For towels that require greater strength, kraft or high-yield pulps may be preferred. With the pollution problems faced by the sulfite process, there has been a shift to kraft and recycled fibers for tissue and towel grades. In general, the brown-colored towels can be assumed to be either kraft or high-yield pulps, and the white towels and tissue bleached kraft, sulfite or recycled fibers. It is difficult to detect the use of recycled fibers in these grades, since fairly white, clean waste grades are used and impurities are removed in the mill. Hardwood fibers have been mentioned in earlier chapters as having the ability to contribute bulk and softness to paper, but they are not entirely satisfactory for use in these grades. For lightweight tissue the small hardwood fibers would not be retained well on the wire and would be lost during the formation of the web.

STOCK PREP

Fibers require refining to develop the bonding ability needed to make strong paper. Bonding, however, creates a more dense, brittle and generally less soft sheet. Accordingly, the tissue and toweling manufacturer must balance strength requirements against softness requirements. Lightly refined softwood fibers can satisfy both requirements. Wet-strength agents, which can help the tissue or towel retain strength when wet, are an obvious need for towels and facial tissue. Urea and melamine formaldehyde have been used in the past, but the need to remove formaldehyde from the environment has led to its replacement by other chemicals, such as amino polyamide-epichlorohydrin resin.

PAPERMAKING OPERATIONS

Formers

Fourdrinier machines are the obvious choice for these lightweight grades. The forming section will tend to be shorter than normal for printing or writing grades, since the stock tends to drain faster. Fourdrinier formers are represented by the drawing in Figure 11.1, which shows a large breast roll with the stock delivered to the wire above the breast roll. The large amount of suction developed on the back of the breast roll can be used to drain the web quickly. This same high-speed forming can benefit the web by freezing the turbulent flow coming from the headbox, or it can hurt the web by removing too many fibers and fines from the web. The exact location of the point at which the stock is delivered to the wire varies from mill to mill, both in the tissue business and on all fourdrinier formers. Drainage on the wire is assisted by either large table rolls or foils. These machines normally operate at high speeds, where foils can be very effective. The web is lifted from the wire by a pickup felt since the web is too light and fragile to be able to sustain a free draw at this point. The same pickup felt can be seen delivering the web to the yankee dryer. Machines of this type can also be used to form toweling, but the delivery from the headbox would likely be shifted to just past the breast roll and the heavier weight might require a longer wire and forming section.

Figures 11.2 and 11.3 show two twin-wire concepts in which the web is formed around a solid forming roll. These formers are normally used only to form tissue, but could conceivably be used for lightweight napkin or towel stock. In Figure 11.2 the stock is delivered to the top of the bottom wire and the water is forced out by passing the web and wire around a solid forming roll. The forming roll

Figure 11.1. Fourdrinier type tissue machine

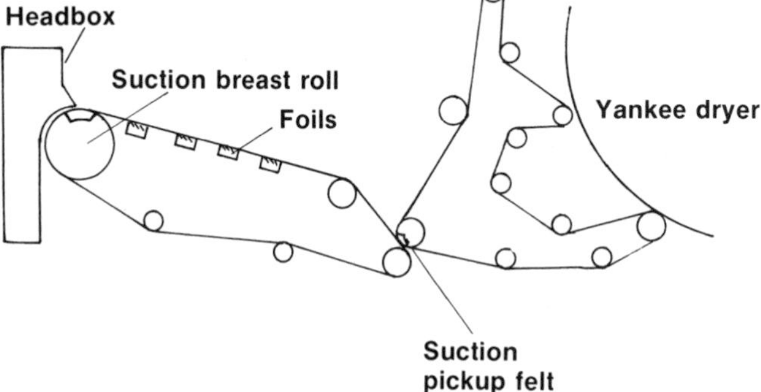

Toilet tissue rewinding machine. A 99-in.-wide roll of tissue is rewound into logs which are accumulated in the large frame then fed to the left to be cut into consumer rolls and packaged.

operate in a continuous manner in each machine reel. In other words, the machine is stopped only to load machine reels or for web breaks. The cores are made, fed to a turret winder where they are rotated into position to have tissue wound on them until they are of the proper diameter, and then rotated to another position to be ejected from the machine while an empty core is inserted.

Folded Products

Facial tissues are generally slit into webs about 8 in. wide, the width of one tissue, and rewound into smaller rolls. These rolls of paper are then loaded into a machine that will fold and box them. The machine is equipped with a large unwind panel, which can accept about 50 of the rolls of tissue from the rewinder. The rolls are positioned such that as they are unwound, the webs can be folded and stacked on top of one another, as indicated in Figure 11.7. Two different types of fold are commonly used: the C fold and the interlocking fold. The interlocking fold is needed for the pop-up box, where each tissue is folded in half and interlocked with the tissue above and below it. As one tissue is removed, the overlapping or interlocking flaps cause the top flap of the next tissue to be pulled up.

The number of rolls is a function of or determining factor for the number of sheets in a box. The webs are folded and stacked into a long log, which is cut into short sections to fit into the boxes. If 50 rolls are fed together, two stacks must be combined to make a box of 100, and so forth. The machine is installed next to

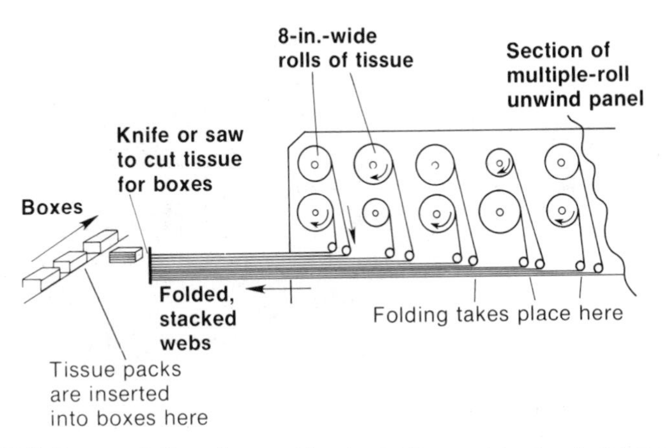

Figure 11.7. Representation of a machine for folding and boxing facial tissue

and operated in coordination with a packaging machine, which accepts folded boxes, opens them, inserts the tissues, seals the box and loads them into a corrugated container. Again we have a continuous operation that converts the rewound rolls into a consumer product ready to ship.

Machine to fold and stack seven webs of c/fold towels at Tagsons Papers Inc., Menands, New York.

Napkin folding machine capable of folding and delivering 5,200 napkins per minute. Rolls are unwound from the left, fed to the right where they are folded with plow folders at the top and fed down to roll folders for folding in the other direction. The double-folded napkins are then cut, stacked, counted and delivered for packaging.

Towels can be processed in a manner quite similar to facial tissue. The c/fold towel machine shown below accepts seven rolls of paper (out of sight behind the machine), which come under the bottom and are folded double with a plow folder. As the folded webs travel up and to the rear, both edges are folded under into the c/fold with the folds on the bottom and then stacked as they proceed to the rear. The stacked log can then be cut and stacked further to the desired number of towels.

Napkins can be converted and packaged in a similar way, but are normally processed only one or two rolls at a time. The tissue and towel machines just described unwind and fold the webs in the machine direction. Napkins are more likely to be unwound in the machine direction and folded across the machine direction, or even in both directions. The parent reel from the paper machine is rewound into rolls that are normally wide enough to produce two or more napkins side by side. These smaller rolls are fed into a machine either singly or, if the product is to be a two-ply towel or napkin, in pairs. The web is cut into sheets and folded into a C or an interlocking pattern, stacked in a delivery pile, counted and then slit by a bandsaw into separate bundles of napkins or towels and banded or boxed. Again the process is continuous except for interruptions to load new rolls.

Embossed napkins can be processed in exactly the same way with the addition of an embosser ahead of the cutting and folding operations. Fancy napkins are likely to be cut from rolls that are only wide enough to make one napkin at a time. The folding of fancy napkins is also more complicated since they must be folded in two directions.

DRY-FORMED PRODUCTS

One notable exception to the manufacturing operations presented so far is the manufacture of paper diapers and sanitary napkins. The bulky, fibrous portion of

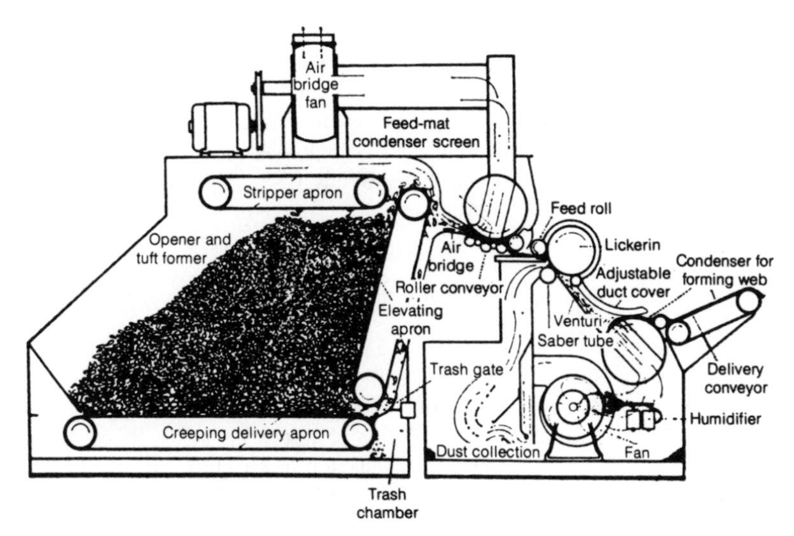

Figure 11.8. Random dry-web process developed by Curlator Corp.

these products is commonly *dry formed*, meaning that the fibers are beaten apart and suspended in a stream of air to be formed into a web. The fibers are deposited on a moving wire similar to the fourdrinier wire, but in a totally different machine (Figure 11.8). The use of air prevents bonding and the accompanying densification of the web, but favors the bulky, absorbent qualities essential in these products. Bonding is not needed since these webs are combined with liners that have the necessary strength and permeability to liquids. A typical diaper converting arrangement is shown in Figure 11.9.

Another recent exception is the manufacture of kitchen towels or industrial wipers by a dry-forming process. The fibers are generally dried after pulping by a fluff drying method. This technique simply squeezes the fibers as dry as possible

Figure 11.9. Typical diaper converting arrangement

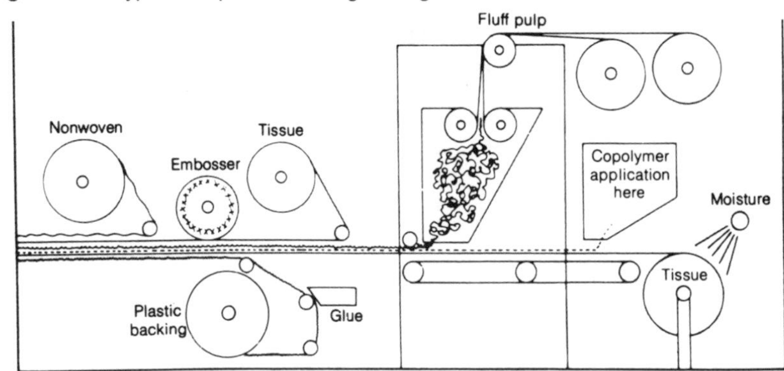

and then dries them with hot air while mixing them and keeping them separated. The dry fibers can then be resuspended in air and deposited on a forming wire. The fibers can be bonded together by the addition of an adhesive to the suspension before or after forming. Adhesive can also be printed onto the formed web at points in a pattern, giving the web a checkered look. Use of a water-resistant adhesive, and the possible inclusion of nonwoody fibers, can make the webs very water resistant and yet extremely absorbent.

SPECIAL PROPERTIES AND TESTS

Probably the most important property of tissue and related grades—and the most difficult to test for—is softness. We all know whether a tissue is soft to our nose or skin when we touch it, but this same sensation is not easy to simulate with a piece of test equipment. The sensation of softness is generally conceded to be a combination of surface smoothness and lack of stiffness. Drape, the fabric industry term for the opposite of stiffness, is possible to observe but not easy to quantify. Smoothness of paper is frequently measured by using air leak instruments, but these are rendered useless by porous tissue. We end up with a machine that draws a folded piece of tissue through a slot, a test that should be affected by the combination of drape and smoothness of the surface. Unfortunately, the final test of softness remains with the consumer, as attested to by television commercials that demonstrate the gentleness or squeezability of the product.

Water absorbency is another property that can be difficult to assess. One aspect of absorbency is the speed with which liquid is absorbed. Another is the total volume of liquid that can be absorbed. Each parameter can be valuable, and can be evaluated by test methods designed to simulate the end use, meaning there are a number of tests aimed at different uses.

Wet strength is easier to measure once we determine what strength property we are interested in. Normally the paper sample is wetted for a specific time period, blotted dry and tested for its tensile strength. The tensile strength may not be truly representative of the type of force to which the product will be subjected during use, but it is a good evaluation of the degree of bonding in the sheet.

REFERENCES

Bambrick, T., and New, W., *Pulp & Paper*, February 1979, p. 104.

Hanson, J. P., "What's going on in non-wovens," *Pulp & Paper*, October 1977, p. 97.

Hanson, J. P., "What's going on in tissue," *Pulp & Paper*, November 1977, pp. 93-102.

12 Cut Size, Bond, Copy Paper and Computer Paper

The grades grouped in this chapter are diverse, yet have some similarities. There are definite similarities in manufacturing operations, but although one mill may make several of these grades, it is unlikely that one mill will make all of them. They are similar in that they are all used in offices and may be purchased or sold by the same people. Accordingly, many are sold through a converter, who purchases rolls of paper from mills and converts those rolls into these products. Some of the grades are more likely to be converted at the mill and shipped either to the jobber or to a supply house where the large lots are split into smaller lots for the convenience of the final customer. Who does what to the paper is a complex decision based on a number of factors and will differ from mill to mill.

With respect to manufacturing operations, it should be assumed that the following factors apply for all of the grades unless otherwise noted. Most are made from a mixture that is primarily bleached softwood kraft. The rest of the furnish will be made up of bleached hardwood kraft and/or secondary fibers. The secondary fibers are likely to be deinked unless exceptionally clean clippings can be obtained from some other converting operation. Internal sizing will be common to all, as will the use of pigment fillers in those grades that need opacity and good printing properties. The paper machine will normally be a flat-wire fourdrinier, using normal pressing and drying sections and surface sizing. Web modifications can be nonexistent or of major importance and will be discussed as needed. All grades will be assumed to be sized with starch at the size press.

CUT SIZE

Manufacturing Operations

Cut size is a rather loose term from the standpoint of the actual paper grades represented. All that is needed is that the paper be trimmed to 16 x 21 in. or less. Typing, writing, tablet, notebook, and bond papers all belong to this group. The only grades to depart greatly from the pulping norms stated earlier are the bonds and watermarked papers. The term *bond* does not signal any divergence from those norms until we expand it by referring to *rag bond*, *cotton bond* or *sulfite bond*. Rag and cotton bond should both contain cotton fibers, which are more pure than wood fibers and make a paper with far better permanence and resistance to abuse. Money paper, stamp paper, and paper used to print stock and bond certificates have need for these properties. The rag or cotton content of these papers is generally rigidly enforced and stated as part of the name or grade. With the strong competition for cotton fibers from clothing manufacture and the reduced availability of 100% cotton scraps, cotton bond has become more expensive and scarce. Sulfite bond has been made as a substitute because of the purity and improved permanence of sulfite pulp over kraft, but sulfite is still not nearly as permanent as cotton, and has also become scarce in response to environmental concerns.

One common manufacturing operation used on much of the bond grades is the *watermark*. To produce watermarked paper, a special dandy roll is needed. The dandy roll is modified by the addition of raised portions on its surface. The older method was to solder wires to the surface of the dandy roll in the pattern desired. This operation is extremely time consuming and expensive and has been replaced by an embossing process. Before the dandy wire is wrapped around the roll, the pattern is stamped into the wire, giving it raised portions where the watermark should be. As this watermarking dandy presses into the wet web on the fourdrinier wire, the fibers are forced aside and the web is made slightly thinner in the pattern of the watermark. Because both the making of the watermarking dandy and its use on the paper machine are costly, watermarked papers are expensive and demand is diminishing.

Another possible departure from the norms listed would be for *erasable bond or typing papers*. This paper either receives a size press application of a bonding agent stronger than starch or it is coated with a release agent. The two mechanisms used to make paper erasable are (1) to improve the surface strength so that the paper can resist the rubbing action of an erasure and (2) to apply a coating that will hold the ink on the surface of the paper. Either mechanism can work, but the latter can run into problems from ink smudging, or use of ink that is not compatible with the coating. A combination of both treatments is the best solution and can be accomplished with gums or polyvinyl alcohol coatings containing fluorocarbon release agents.

Converting Operations

Whether converted in the mill or at a converting plant, the same machinery and operations are used for cut size grades. The rolls produced on the paper machines are first rewound and slit into narrower rolls. The rolls are then loaded into machines similar to the one shown in Figure 12.1, which will process them into finished products, including wrapping and packaging in shipping cartons. Older machinery that separates some of these operations may still be used, but the trend is toward the faster, more economical machines that do everything.

It can be seen in Figure 12.1 that multiple rolls are fed to a device called a *folio sheeter*. This device slits the rolls in the machine direction first, then feeds these webs into a rotary fly knife to cut them across the machine direction into individual sheets. The sheeter in Figure 12.1 shows the rolls being cut into four sheets across the width of the rolls. The cut sheets are stacked on top of one another and held until the desired number is collected; then the stacks are fed to the next operation.

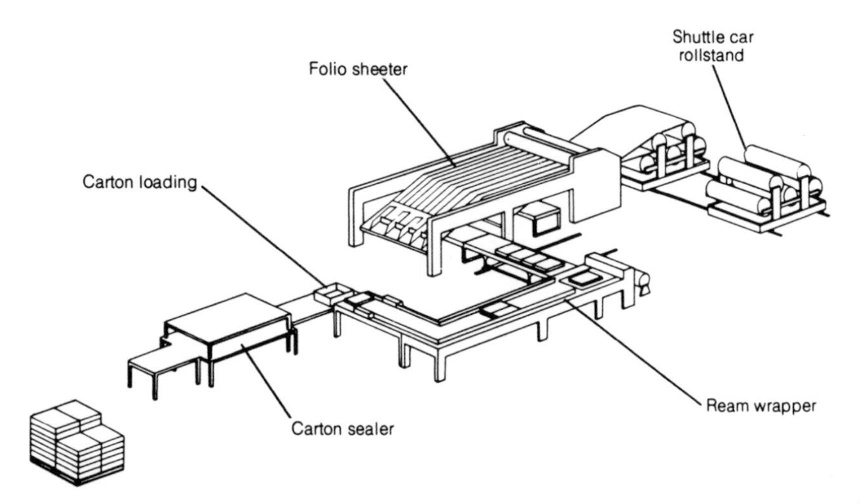

Figure 12.1. Schematic drawing of a cut-size operation

Other machines are used that differ from the one shown in Figure 12.1 primarily in this first part of the overall operation. The major difference is in the sequence of cutting operations. The oldest manual methods cut the rolls into large sheets and then use guillotine trimmers to cut the large sheets into the desired sizes. Other automatic machines cut the rolls across the machine direction with the fly knife into wide sheets, which are then cut in the machine direction to the finished size. Either of these machines can be modified to accomplish other tasks as well.

If the product is to be ruled or lined, the ruling operation generally comes first. The lines are applied by a roll that has raised bars in the pattern desired. The ink is a water-based dye that strikes into the paper and dries immediately. Multiple rolls are normally fed into these machines such that each web is ruled and then they are fed to the fly knife together to be cut into sheets with the long dimension across the machine direction and therefore across the width of the converting machine. The wide sheets are stacked on top of one another until enough sheets are collected for the product being made, and then the sheets move ahead to the following operations.

If the product requires that holes be punched in the sheets, the punching can be performed either on the combined webs just ahead of the sheeter or on the stacked sheets. The location of this operation depends on the need to register the holes with the cut, the need for clean punched holes and the capability to drill holes through 100 or more sheets in a pile. In short, punching and drilling are done in different places for different products. The stacked sheets must next be cut into sheets of the proper size for the finished product. This cut is made by guillotine type cutter knives, which are able to cut the wide sheet into sets of finished sheets simultaneously.

The cut sheets can be stacked into larger piles; spacer sheets can be inserted; top and bottom tablet covers can be inserted; glue, tape or wire binders can be attached; and the final product wrapped or boxed. Essentially the same machine can be used for a wide variety of products by adding or omitting different capabilities. Among the products converted with this type of machine are: typing paper, plain paper, copier paper, lined loose-leaf binder paper, spiral notebooks and tablets.

COMPUTER PAPERS

There are two grades of *computer papers* to be considered here: *cards* and *paper for high-speed printers*. The cards must have stiffness and excellent dimensional stability to be able to be processed well. Stiffness also improves their ability to be punched cleanly. Long softwood fibers are generally desirable, along with starch internally and at the size press to help develop stiffness and dimensional stability.

Paper for high-speed printers is frequently groundwood containing, made on fourdrinier or twin-wire machines, size press coated and calendered. Conversion includes printing of light and dark bars across the web (ruling), punching holes on the edges of the web, perforating and folding into the boxes from which the paper will be fed into the printers. The paper is first rewound off the paper machine and trimmed into webs of the desired final width. The printing is commonly done with narrow web offset printers that may use direct litho plates. If the product is to be a multiple web with carbon paper interleafed, the press must have multiple printing stations so the webs can be printed simultaneously; the carbon paper web interleafed; and then the punching, perforating, folding and packaging performed. Single webs are processed separately, although the press may be equipped to handle several webs, processing and packaging each one separately yet simultaneously with the others. Such a press could be designed to be quickly converted to handle multiple-web products with interleafed sheets.

COPY PAPERS

Waxed or Carbon Papers

Carbon papers are the oldest form of office copy paper and are essentially made by the same process that has been used for years. The paper needs to be thin and smooth, especially if multiple copies are desired, as in the case of preassembled manifolds. Softwood kraft is essential to give the web the strength needed at the light basis weights produced. Yankee dryers are sometimes used to make machine-glazed paper because of the smoother surface they produce. Calendering is needed for non-glazed papers to help develop smoothness and density.

The carbon coating is applied in a wax, which adheres to the surface of the web without penetrating far enough to mark the back side of the sheet above it during use. The wax coating is applied to wide webs by rod or air knife coaters, and the web is then rewound and slit into the narrower webs needed for use in manifold or business-form–making.

Carbonless and No-Carbon Grades

The carbon-coated papers are potentially messy in the office and can cause problems in the recycling of wastepaper. These problems led to the development of special grades that can produce copies without carbon. These operations rely on the use of two different chemicals that are colorless when separate but develop a color when mixed. They are applied to the webs in different manners based on the particular process used. All rely on a process known as *microencapsulation*. One of the two chemicals is encapsulated into small spheres formed of a fragile shell. These capsules are coated onto the backside of one sheet, and the pigment onto which the color develops is coated onto the front side of what will be the copy. Typing, writing with a hard pen or pencil or any pressure will fracture the capsules, allowing the two to contact each other and the color to develop. The obvious problem with this method is that the two papers must be used together or no copy will result. Both chemicals can be applied to the same side of a sheet, but the potential for accidental or unintentional color development is greater. This paper can be used behind any original—or develop marks when subjected to any pressure or impact.

All copy papers discussed so far are sensitive to pressure in their processing. Winders used with these papers are generally center-shaft rather than surface winders. A surface winder, suitable for other grades, winds the roll by pressing the core against one or two other rolls, which are driven and feed the paper onto the core. To wind paper onto a core that is driven and yet keep the paper from contact with any other rolls, the rotational speed of the core must be changed to keep the circumference of the roll operation at the speed of the machine. These machines are more difficult to control and are not always as able to build large, solid rolls; therefore they are limited to applications where their special characteristics are required.

The carbonless copy papers have run into problems with pollution due to the chemicals used in the encapsulation process. Therefore, we find that each type of process has advantages over the other and both are used.

Plain Paper Copiers

The basis for the plain paper copier is the use of a semiconductor drum or belt. This is the process developed by Xerox and therefore also known as *xerography*. The drum or belt is charged over its entire surface, then exposed with an image of the original. The charge is retained in the dark areas of the original and discharged in the areas exposed to light. The charge is then used to pick up oppositely charged ink particles. The ink or resin particles are then transferred to the paper by pressing them against the paper and discharging the carrier. The ink is dried or the resin particles are fused with heat and the copy is finished. These copiers are called *plain paper copiers* since the paper used in them need not have special electrical properties. The paper must, however, have heat stability if it is to be used in a copier that fuses the resin with heat. The main problem is curl induced by the drying of one side of the sheet and not the other during the fusing process. The paper first must lie flat on the feed pile of the copier so it can be fed in.

Papers for these copiers generally have a high ash content or filler level to help reduce the curl and make the paper lie flat. Hardwoods and recycled fibers are also used along with a basic level of softwoods to give the sheets strength.

This process is now being used in *laser printers*, where the laser is used to develop the image. The laser allows the copy to be generated by computer or received in a digitized form from any source. They are either being used for very high-quality output on smooth paper or for extremely high-speed, medium- to low-quality output on lower-quality, fan-fold computer paper or business forms.

Electrostatic Copy Papers

In the electrostatic copy machines, the same charging, discharging and imaging processes are used, but the paper becomes an active, rather than a passive, participant in the operation. *Electrostatic copy paper* is coated with a semiconductive coating containing zinc oxide as the pigment particle that can take the charge. The zinc oxide must be held in the coating by a nonconductive adhesive or resin so that the charge in the image areas can be retained and not diffused, making a fuzzy copy. Furthermore, the paper that carries the coating must either be sufficiently nonconductive or shielded from the charged pigment particles. The paper may therefore be coated with a precoat to insulate the electrographic coating from the paper and to keep it on the surface. Even if the conductivity of the paper can be kept low enough to be no problem, a precoat may still be necessary to keep the electrographic coating on the surface. The performance of this coating is sensitive to its thickness and since the surface of the web is irregular, the coating thickness can vary. Smooth-surfaced paper is obviously desirable for these grades. The smoothness of the surface can be improved by precoating. The electrographic coating is preferably applied with an air knife coater since it applies heavy coats of uniform thickness. A blade coater is better for the precoat since it tends to fill in the valleys and create a smoother surface.

The electrostatic copiers have diminished in popularity due to higher costs and a general dislike for the look and feel of the paper. This paper was the original version of *electrofax* or *facsimile* transmissions, which have increased dramatically in popularity. Although this paper was the original, the newer, cheaper *fax* machines are more likely to use *thermal copy* or *thermofax* paper. This paper is coated with a pigment and chemical mixture that will become black or dark brown when heated. The copy is produced by passing the sheet in contact with a print head, which contains pins that can be heated on demand. The original can be generated by computer or transmitted to the copier in a digitized form to create the copy. This paper has also been used for copy machines, where the copy was produced by placing the original and copy sheet together and exposing them to a bright light that would heat the type on the original, but only if it contained carbon or was dense enough. Besides the obvious problems with having the proper original, and potential for loss of the copy due to exposure to heat, the process did not have the good appearance or permanence of xerography and is no longer popular as a copy process.

Glossary

ABRASION RESISTANCE. Ability of a paper product to withstand abrasion. Measured by determining degree and rate that a sample loses weight under specific rubbing action of an abrading substance, such as an eraser.

ABSORBENCY. Property of pulp, paper, and its constituents and products that permits the entrainment and retention of other materials it contacts, such as liquid, gaseous, and solid substances.

AIR DRY (a.d.). Weight of moisture-free pulp or paper plus 10% moisture based on traditional assumption that this amount of moisture exists when they come into equilibrium with the atmosphere.

ALUM. Papermaking chemical commonly used for precipitating rosin size onto pulp fibers to impart water-resistant properties (when used for water treatment) to the paper. Also called *aluminum sulfate.*

ANTITARNISH PAPER. Term originally applied to higher-weight tissues used for wrapping silverware, but now used for all papers so prepared that they will not rust or discolor razor blades, needles, silverware, etc.

APPARENT DENSITY. Weight (mass) per unit volume of a sheet of paper obtained by dividing the basis weight by the caliper (thickness).

BASESTOCK. Paper or board to be further treated in various ways.

BASIS WEIGHT. Weight in pounds of a ream of paper, usually consisting of 480, 500, or 1,000 sheets of a specified size. Also expressed as g/m^2 (grammage).

BEATER. Large, longitudinally partitioned, oval tub used to mix and mechanically "work" pulp with other ingredients to make paper.

BEATING. See *refining.*

BINDERS BOARD. Gray-colored, glazed board often used in the binding of hardcover books.

BIOLOGICAL OXYGEN DEMAND (BOD). Amount of dissolved oxygen consumed in five days by biological processes breaking down organic matter in mill effluent.

BISULFITE PULP. Pulp made by the bisulfite cooking process using bisulfite cooking liquor.

BOGUS. Product that has been made from wastepaper or other inferior materials to imitate higher-quality grades.

BOND. Class of printing/writing papers made from bleached chemical woodpulps and cotton fibers.

BONDING STRENGTH. Intralayer binding force in a multi-ply paperboard or laminate. Also refers to the degree of adherence of coating and film on a sheet and to the interfiber binding force within a sheet.

BONE DRY (b.d.). (1) Descriptive term for the moisture-free conditions of pulp paper. See *oven dry.* (2) Refers to air containing no vapor.

Source: Excerpted from *Pulp & Paper Dictionary* by John R. Lavigne (San Francisco, Miller Freeman Publications, 1986).

BOOK PAPER. Paper suitable for printing and other uses in the graphic arts industry.

BRIGHTNESS. Measure of the degree of reflectivity of a sheet of pulp or paper for blue light measured under specified standard conditions. Also called *whiteness*.

BRISTOLS. Heavy-grade papers possessing higher-than-average quality characteristics. Rigid enough to be used for announcements, invitations, postcards, etc.

BROKE. Paper trimmings or damaged paper due to breaks on paper machine and in finishing operations.

BULK. Compactness property of a sheet in relation to its weight (whose value decreases as compactness increases).

BURSTING STRENGTH. Resistance of paper to rupture when pressure is applied to a side by a specified instrument. Also called *burst*, *mullen*, and *pop strength*.

CALENDER. Piece of processing equipment located at dry end of paper machine, consisting of a set of rolls through which paper sheet is passed for smoothing, leveling, and gloss improvement.

CALENDER FINISHED. Paper and paperboard passed through a calender to improve surface characteristics by application of pressure, friction, or moisture.

CALIPER. Thickness of a sheet of paper or paperboard, measured under certain specifically stated conditions, expressed in units of thousandths of an inch (called "mils" when referring to paper, "points" when referring to paperboard). Also called *thickness*.

CARBONIZING PAPER. Lightweight, uncoated paper made from unbleached chemical and/or mechanical pulps and surface-coated with a carbon solvent or wax so that it takes up carbon inks and releases them under pressure, thereby duplicating the inked areas being printed.

CARBONLESS PAPER. Copying paper that is treated or coated so it can be used without needing carbon coating or interleaved carbon paper.

CARLOAD. Quantity of paper shipped from mill in one freight car. Must exceed a freight classification zone minimum weight to qualify for carload freight rate.

CAST-COATED PAPER. Very high-gloss coated paper and paperboard with surface characteristics produced by allowing applied coating to harden while in contact with surface of steam-heated, highly polished, chrome-plated drum.

CELLULOSE. Chief substance in cell walls of plants used to manufacture pulp.

CELLULOSE WADDING. See *wadding*.

CHEMICAL PULP. Mass of fibers resulting from reduction of wood or other fibrous raw material into component parts during cooking phases with various chemical liquors in processes such as sulfate, sulfite, soda, and NSCC.

CHEMI-THERMOMECHANICAL PULP (CTMP). Pulp made by thermomechanical process in which woodchips are pretreated with a chemical, usually sodium sulfite, either prior to or during presteaming as an aid to subsequent mechanical processing in refiners.

CHIPBOARD. Inferior-quality, low-density, solid or lined paperboard made primarily from recycled wastepaper stock and used in low-strength applications.

CLARIFIERS. Storage tanks in which suspended solids are allowed to settle and be removed from green and white liquors in the causticizing areas of a pulp mill.

COATED. Papers and paperboards that contain a layer of coating material, such as clay or pigment, in combination with an adhesive.

COMBINATION BOARD. Multi-layered, cylinder-made paperboard having outer and inner layers made from different pulpstocks.

COMBINED BOARD. Multi-layered board made by uniting a number of boards with proper adhesives.

CONSTRUCTION PAPER. Heavy type of paper used for watercolor and crayon artwork, made in various colors primarily from groundwood pulp.

CONTAINERBOARD. Single- and multi-ply, solid, and corrugated boards used to make boxes and other containers for shipping materials.

COOK. Process of reacting fiber-containing materials, such as wood, rag, straw, and bagasse, with suitable chemicals, usually under high temperature and pressure, in order to reduce them into component parts so that acceptable fibers can be separated and made into pulp.

CORD. Pulpwood volume measurement indicating a pile measuring 4 ft x 4 ft x 8 ft, equaling 128 ft^3 (3.62 m^3).

CORRUGATED BOARD. (1) A pasted, single- or double-faced, multi-layered board having a fluted bottom or middle layer. (2) The fluted paperboard after it has gone through the corrugating operation and before it is pasted to the flat facing board sheet.

CORRUGATING MEDIUM. Paperboard made from chemical and semichemical pulps, sometimes mixed with straw or recycled paperstock, that is to be converted to a corrugated board by passing it through a corrugating machine.

COVER PAPER. Paper used as a protective covering for books, pamphlets, magazines, catalogs, and boxes.

CREPE PAPER. Low-basis-weight paper made from sulfite, sulfate, or mechanical pulp and given a simulated crinkly finish by crowding the web sheet over a roll with a doctor blade.

CROSS DIRECTION (CD). Side-to-side direction of a paper machine or the paper sheet made on it, as opposed to *machine direction*, which runs from head to exit end.

CUNIT. Pulpwood measurement equivalent to a volume containing 100 ft^3 solid of unpeeled wood.

CURL. Paper or paperboard deformation caused by nonuniform distribution of strains and stresses throughout the sheet as a result of uneven internal moisture and conditioning.

CUT SIZE. Fine paper cut to specific end-use dimensions (16 in. x 21 in. or less) on a paper trimmer usually of the guillotine or rotary type.

CYLINDER MACHINE. Machine primarily used to make paperboards. The forming cylinders are covered with wire so that, as they turn within a vat filled with stock solution, fibers are picked up to form a web on the surface with water draining through and passing out at the ends. The wet sheet is then transferred off the cylinder onto a felt for possible combining with other sheets (multiple cylinders on same machine) and subsequent pressing and drying.

DEINKING. Removal of ink and other undesirable materials from wastepaper by mechanical disintegration, chemical treatment, washing, and bleaching before reusing as a source of papermaking fiber.

DIE CUT. Paper and paperboard products cut by a metallic die to specified dimensions or form.

DIGESTER. Pressure vessel used to chemically treat chips and other cellulosic fibrous materials such as straw, bagasse, rags, etc., under elevated temperature and pressure in order to separate fibers and produce pulp.

DIMENSIONAL STABILITY. Ability of a sheet to maintain its original machine and cross-machine dimensions with time and under variable moisture and relative-humidity conditions.

DRY END. Portion of a paper machine where sheet moisture is removed by evaporation. Consists of several dryer sections or air dryers, depending on type and size of the machine.

DRYERS. Portion of a paper machine where water is removed from wet paper by passing it over rotating, steam-heated, cylindrical, metal drums or by running it through a hot air stream.

DUPLEX. (1) Papers and paperboards with each side having a different color, finish, or surface texture, which is produced on the paper machine or by pasting. (2) General term referring to multi-ply paper and paperboard and to bags made of two separate sheets of paper.

ELECTROSTATIC COPY PAPER. Smooth-finished, stable, medium-weight bond paper made from chemical pulps. Generally treated with a zinc oxide coating material and used on dry-type office copying machines.

ELMENDORF TEST. Test commonly used in paper mill laboratories to determine tear-resistant property of paper. Also called *tear test.*

ELONGATION. Physical property of a paper sheet that allows it to experience a certain degree of stretching.

EMBOSSED. Paper finish obtained by mechanically impressing a design on the sheet with engraved metallic rolls or plates.

ENAMEL. Clay coating on coated paper.

ENGLISH FINISH. Medium finish applied to a sheet of paper that is smoother than the finish coming off dryers or calenders, but not as smooth as the finish coming off supercalenders.

FABRIC PRESS. Paper machine wet press that uses a special multiple-weave fabric belt sandwiched between the regular felt and the rubber-covered roll, increasing the capacity to receive and remove water from the nip between rolls.

FEEL. Evaluation of paper surface finish by sense of touch.

FELT. Woven belt of wool, cotton, or synthetic fibers used to transport sheet of paper between rolls of press section (wet felt) and against dryer drum (dryer felt) in dry-end section of paper machine.

FELT SIDE. Top side of paper sheet as it is formed on wire of wet end of paper machine, which later comes in contact with felts during subsequent drying phases.

FIBER. Elongated, tapering, thick-walled cellular unit that is the structural component of woody plants.

FILLER. (1) Substance added to pulpstock to fill spaces between fibers and enhance printing properties of paper made from it. (2) Inner layers of multi-ply paperboards.

FINE PAPERS. High-quality printing/writing and cover papers having excellent surface characteristics for pen and ink writing.

FINISH. Surface characteristics of sheet of paper, such as smoothness, appearance, and gloss, as determined visually.

FINISHING. Processing of paper after completion of papermaking operations, including supercalendering, slitting, rewinding, trimming, sorting, counting, and packaging prior to shipment from mill.

FLEXOGRAPHIC PRINTING. Rotary letterpress printing process using ink made of aniline dyes and pigments (mixed with a binder) that dries primarily by evaporation due to the solvent vehicle (with rapid evaporation properties) used. Also called *aniline printing.*

FLUFF PULP. Thick sheet or batt of woodpulp fibers manufactured in roll or bale form and suitable for dry disintegration into individual fibers.

FLUORESCENT PAPER. White paper made with synthetic dyes that produce a brighter appearance by increasing the ability to reflect light, or paper that is surface-coated with colored, light-emitting dyestuff materials (which reflect white light as a color).

FORMATION. Physical distribution and orientation of fibers and other solid constituents in the structure of a sheet of paper that affects its appearance and other physical properties.

FOURDRINIER WIRE. Continuously traveling, endless, woven, metallic, or plastic screen belt located in wet-end section of fourdrinier paper machine. Pulpstock is fed onto wire so that water is drained from it as fibers become oriented to form a continuous web.

FREENESS. Ability of pulp and water mixture to release or retain water on drainage.

FREE-SHEET. Sheet of paper containing no mechanical pulp fibers or made of pulp subjected to minimal refining or hydration, which allows water to drain quickly when sheet is formed on fourdrinier wire. Also called *woodfree*.

FURNISH. Various pulps, dyes, additives, and other chemicals blended together in stock preparation area of paper mill and fed to wet end of paper machine to make paper or paperboard. Also called *stock*.

GLASSINE. Light, dense, translucent paper made from highly refined chemical pulp and possessing a high degree of hydration. Used as envelope windows and in protective packaging for foodstuffs, candy, tobacco products, chemicals, and metallic items.

GLOSS. Property of paper sheet surface that produces a shiny, highly reflective appearance when light is reflected from it.

GRAIN. Directional alignment of fibers in a paper sheet structure.

GRAMMAGE. See *basis weight*.

GRAVURE PRINTING. Intaglio printing process employing minute engraved "wells." Generally, deeply etched wells carry more ink than a raised surface; hence, they print darker values. Shallow wells are used to print light values. A doctor blade wipes excess ink from the cylindrical printing surface. Rotogravure employs etched cylinders and web-fed stock. Sheet-fed gravure, as its name implies, involves individual sheet feeding.

GROUNDWOOD PAPER. Paper that is made from a furnish containing a large percentage of groundwood pulp.

GROUNDWOOD PULP. Slurry produced by mechanically abrading fibers from barked logs through forced contact with the surface of a revolving grindstone. Used extensively to make newsprint and publication papers.

HARD-SIZED. Paper and paperboard made resistant to water and ink penetration by exposure to high degree of sizing treatments.

HARDWOOD. Pulpwood from broad-leaved dicotyledonous deciduous trees.

HEMICELLULOSE. Alkali-soluble, noncellulosic polysaccharide portion of a wood cell wall.

HOLDOUT. Ability of paper or board to resist surface liquid penetration.

HOTMELTS. Plastic or wax coating materials in a molten state applied to paper or paperboard sheet to produce fluid-resistant surface with high gloss.

INDUSTRIAL PAPERS. Paper made for purposes such as industrial packaging, tissues, wrappings, impregnating, insulating, etc.

INTEGRATED MILL. Mill manufacturing complex in which all pulp and papermaking operations are conducted at one site.

INTERNAL BONDING STRENGTH. See *bonding strength*.

JOB LOT. Out-of-specification, defective, or discontinued types of paper made in small quantities for special orders and sometimes sold at lower-than-standard prices.

JUMBO ROLL. (1) Larger-than-normal roll of paper as it is slit and rewound. (2) Roll of paper usually greater than 12 in. in diameter and used for converting into user products.

KRAFT PAPER. High-strength paper made from sulfate pulp, usually with a naturally brown color from unbleached pulp. Also called *sulfate paper*.

KRAFT PULP. Fibrous material used in pulp, paper, and paperboard manufacturing, produced by chemically reducing woodchips into their component parts by cooking in a vessel under pressure using an alkaline cooking liquor. Also called *sulfate pulp*.

LACQUER. Organic solution with volatile solvents used for coating paper to give high surface gloss, grease resistance, heat sealing, and improved surface appearance.

LAMINATED PAPER. Multi-ply paper and paperboard consisting of firmly united, superposed layers, which may be bonded with resin or adhesive.

LEDGER PAPER. Strong, highly sized paper made from bleached chemical woodpulp. Used to make accounting and record books; also used with accounting machines.

LETTERPRESS PRINTING. Printing process in which ink is applied to paper, paperboard, or film from raised portions of printing plates or type.

LIGNIN. Brown organic substance that acts as an interfiber bond in woody materials.

LINERBOARD. Kraft paperboard, generally unbleached, used to line or face corrugated core board (on both sides) to form shipping boxes and various other containers.

MACHINE-COATED. Paper and paperboard that have surface coating of adhesives and minerals applied while being made on the paper machine, as an integral part of the papermaking operation.

MACHINE-DIRECTION (MD). Direction from the wet end to the dry end of a paper machine or to a paper sheet parallel to its forward movement on a paper machine.

MACHINE-FINISH (MF). Surface finish produced on an uncoated sheet of paper as it is being made on the paper machine, and usually accomplished with limited calendering on the machine calender stacks.

MACHINE-GLAZED (MG). High-gloss surface finish produced on the wire side of a sheet by passing it over a large-diameter, highly polished, steam-heated roll as used on yankee-type paper machine dryers.

MANIFOLD PAPER. Very thin regular bond paper with glazed or unglazed finish used to make carbon copies of letters.

MATTE. Paper and surface finishes with very low gloss or luster.

MECHANICAL PULP. Pulp produced by reducing pulpwood logs and chips into their fiber components by the use of mechanical energy, via grinding stones, refiners, etc.

MEDIUM. See *corrugating medium.*

MOISTURE CONTENT (MC). Percent of water by its weight in paper, pulp, paperboard, chips, etc., which will vary according to atmospheric conditions because of the ability of these types of materials to absorb or emit moisture.

MOISTUREPROOF. Ability of paper and paperboard to resist the penetration of water vapor.

MOISTURE VAPOR TRANSMISSION. Ease or rate of water vapor permeation in a sheet of paper.

MOTTLE. (1) Random nonuniformity in printed gloss, or the visual density or color of a printed area caused by uneven absorption of ink by paper. (2) A surface effect produced by the addition of heavily dyed fibers of a different color in the stock furnish.

MULLEN. See *bursting strength.*

MULTI-PLY. Paper or paperboard sheet made up of two or more layers.

NEWSPRINT. Grade of paper, combining high percentages of groundwood pulp, made especially for use in the printing of newspapers.

NINEPOINT. Paperboards with a thickness of 9 mils (0.009 in.) used as the fluted component in the manufacture of multi-ply combined board or wrapping. See *corrugating medium.*

NONWOVEN FABRIC. Sheet of cloth-like material made from long natural and synthetic fibers and formed from a slurry on a wire screen, such as a wet end of a paper machine, or by laying on a fine mesh screen from an air suspension.

OFF-MACHINE COATING. Process of applying coating material to a sheet of paper or paperboard in a location that is away from the machine on which it is made.

OFFSET PRINTING. Process of indirect printing in which an impression of type or a design on a flat plate is printed on a rubber-blanketed cylinder from which it is impressed ("off-set") upon the surface to be printed.

ONIONSKIN. A thin, lightweight, transparent paper made especially for producing typewritten copies of correspondence.

OPACITY. Ability of substances such as paper, flue gases (smoke), and liquids to resist transmission of both diffuse and nondiffuse light through them. Prevents show-through of dark printing in contact with backside of sheet of paper.

OVEN-DRY (o.d.). Moisture-free conditions of pulp and paper. Usually determined by drying a known sample to a constant weight in a completely dry atmosphere at a temperature of 100°C to 105°C (212°F to 221°F). Also called *bone dry*.

PAPER. Homogeneous sheet of felted cellulose fibers, bound together by interweaving and by the use of bonding agents, and made in a variety of types.

PAPERSTOCK. Water slurry of various pulp fibers, dyes, additives, and chemicals that is pumped to the paper machine for forming into a sheet.

PAPETERIE. Chemical pulp and cotton-fiber content paper, made especially for conversion to a class of writing types, usually correspondence stationery.

PATENT COATED. Type of paperboard that is made and lined with white or colored fibers on a multi-cylinder machine, possessing a high surface finish suitable for use in making cartons.

PICK. (1) Phenomenon of pulp or fibers pulling away and sticking to paper machine parts, such as rolls, in the wet and dry end sections. (2) Paper mill control test to determine surface adhesion properties of paper. (3) Small particles of paper that loosen from the surface of paper, especially during printing.

PIGMENT. An insoluble mineral or organic powder used as a dye to color paper and as an additive to impart specific propertiesbulk, porosity, opacityto the sheet.

PLY. Layer that makes up a multi-layered, pasted, or multi-cylinder-formed paperboard.

POINT. Measurement of thickness of a sheet of paper or board (0.001 in.).

POROSITY. Ability of fluids to pass through paper and paperboard, related to size, shape, and distribution orientation of the pores in a sheet and the compactness of the fibers.

PRESSURE-SENSITIVE PAPER. Good-strength paper coated with a pressure-sensitive type of adhesive and converted to tapes and labels.

PRINTABILITY. Ease with which paper can be printed to high-quality standards with the least amount of spoilage.

PULP. Fibrous material produced by mechanically or chemically reducing woody plants into their component parts from which pulp, paper, and paperboard sheets are formed after proper slushing and treatment. Also used for dissolving purposes (dissolving pulp or chemical cellulose) to make rayon, plastics, and other synthetic products. Also called *woodpulp*.

RAG CONTENT. Paper containing from 25% to 75% cotton or rag fibers, including bond, ledger, and specialty papers.

REAM. Stack or package of paper containing a number of sheets (usually 480, 500, or 1,000) designated as standard for that grade.

REFINER MECHANICAL PULP (RMP). Pulp made by processing untreated woodchips in mechanical atmospheric refiners.

REFINING. Pulp and paper mill operations conducted of fiber suspensions to rub, brush, crush, fray, or cut fibers as desired. Imparts such characteristics as increased capacity to absorb water and improved sheet formation.

REGISTER PAPER. Lightweight writing grade made from chemical pulp and possessing good tensile and tearing strength. Comes in rolls especially for automatic register machines and in flat, folded, snap-apart packages for computer printouts and other copying purposes.

RELATIVE HUMIDITY. Actual amount of water vapor present in the air as compared with the maximum amount of water vapor the air could hold at that temperature.

RELEASE PAPER. Type of paper made especially for easily removal from sticky surfaces.

ROSIN. Material made up of a suspension and used for internal sizing of paper and paperboard.

ROTOGRAVURE PRINTING. See *gravure printing*.

RUNABILITY. (1) In the paper mill, how well pulp stock furnish to the paper machine forms a sheet on the wire and passes through the drying and finishing operations. (2) Used by customers in reference to how well the paper performs in their converting operations, such as on printing presses.

SCREENINGS. Rejected materials, such as knots, shives, and large bark particles, from the screening operations of pulp suspensions in a pulp mill.

SECONDARY FIBER. Any type of paper- and paperboard-making fiber obtained from wastepapers and other used, reclaimable fiber sources.

SEMICHEMICAL PULP. Lower-quality pulp made by cooking fibrous materials in a neutral sodium sulfite/sodium carbonate cooking liquor followed by a final separation of the fiber using unpressurized mechanical means.

SHEET. Flat piece of any type of pulp, paper, and paperboard having a variety of characteristics, sizes, and finishes.

SILVICHEMICALS. Chemical byproducts of the woodpulping process and other chemicals derived from wood.

SIZE. Substance such as rosin, gelatins, glues, starch, or waxes added to paperstock furnish or to the surface of a sheet in order to give water-resistant properties.

SIZE PRESS. Paper mill processing unit consisting of two usually rubber-covered rolls located between two dry end sections of the paper machine. Applies size solution to the surface of the paper sheet.

SLURRY. Liquid mixture consisting of suspended fibers, fillers, coating pigments, and other solid material in water or adhesive, used in the papermaking process.

SLUSH. Pulp stockwater suspension thin enough to flow or to pump through a pipeline, usually running about 1% to 6% consistency.

SODA PULP. Pulp made by the cooking of woodchips from deciduous or broadleaf trees in a sodium hydroxide or caustic soda solution.

SOFTWOOD. Wood obtained from evergreen, cone-bearing species of trees, such as pines, spruces, hemlocks, etc., which are characterized by having needles.

SOLID BOARD. Single-ply, homogeneous types of paperboards, made from the same stock throughout the sheet structure.

SPENT LIQUOR. Used cooking liquor in a chemical pulp mill that is separated from the pulp after the cooking process. Contains lignins, resins, carbohydrates, and other substances extracted from the material being cooked.

SPLICE. Joint made in a continuous sheet of paper with glue or adhesive tape when a break occurs in the web during winding or rewinding into a roll.

STANDARD CONDITIONS. Unless otherwise specified, assumed to be 29.92 in. of mercury and 70°F.

STARCH. Type of papermaking adhesive and sizing material made primarily from corn and potatoes. Produces a higher degree of rigidity in a sheet and improves the finish by causing the fibers to lie flat.

STIFFNESS. The ability of paper or paperboard to withstand bending or crushing forces.

STOCK. (1) Fibrous mixture that is made into paper. May consist of one or more types of beaten or refined pulps, with or without suitable fillers, dyes, additives, and other chemicals. Also called *furnish.* (2) Paper suitable for a particular use, such as coating raw stock, milk bottle stock, tag stock, etc.

SULFATE PROCESS. An alkaline pulp manufacturing process in which the active components of the liquor used to cook chips in a pressurized vessel are primarily sodium sulfide and sodium hydroxide, with sodium sulfate and lime being used to replenish these chemicals in recovery operations. Also called *kraft process.*

SULFITE PROCESS. An acid pulp manufacturing process in which chips are reduced to their component parts by cooking in a pressurized vessel using a liquor composed of calcium, sodium magnesium, or ammonia salts of sulfurous acid.

SUPERCALENDER. Auxiliary piece of papermaking equipment used on some paper machines to obtain a denser paper with a higher finish than paper obtained on a calender.

SURFACE-COATED. Any paper or paperboard sheet that has a coating material applied to one or both surfaces.

SURFACE-SIZED. Paper whose surface is treated with a sizing material after the sheet is formed on a paper machine, or in a separate, off-machine operation, occasionally after it has been internally sized.

SYNTHETIC FIBERS. Short filaments that are extruded or spun from synthetic resin materials and used in the manufacture of synthetic paper.

TACK. Sticky property of paper and paperboard adhesive and glue-coating materials.

TALL OIL. Byproduct made from the resins, fatty acids, and soap removed during the evaporation of sulfate black liquor.

TARNISHPROOF PAPER. See *antitarnish paper.*

TEAR STRENGTH. Resistance of a paper sheet to tearing, usually measured by the force required to tear a strip under standardized conditions.

TENSILE STRENGTH. Resistant property of a sheet to pull or stress produced by tension. Expressed as the force per unit width of a sample that is tested to the point of rupture.

TEST LINERBOARD. Types of paperboard that meet specific tests (Rule 41) adopted by the packaging industry to qualify for use as the outer facing layer for corrugated board, from which shipping containers are made.

TEXT PAPER. Good-quality, laid or woven book paper with a medium vellum-like finish.

THERMOMECHANICAL PULP (TMP). Pulp made by presteaming chips and reducing them into their fiber components during an initial mechanical treatment in refiners under elevated temperature and pressure. Subsequent refining done at atmospheric pressure.

TRANSPARENCY. Ability of paper to allow light rays to pass through it in such a manner that objects behind it can be clearly seen.

TRIM. Dimension of the widest sheet of paper that can be made on a paper machine, not including the edges (which are normally cut off).

TWIN-WIRE FORMER. Type of multi-ply paperboard machine having two wires (or fabrics) between which the sheet is formed.

TWO-SIDEDNESS. Visual difference between the top or felt side of a sheet of paper and the bottom or wire side.

VEGETABLE PARCHMENT. Hard, dense sheet of grease-resistant paper, having a very high wet-strength property.

VELLUM FINISH. Smooth, dull finish applied to book and stationery paper surfaces to simulate sheets originally made from young calves' skin.

WADDING. Single- or multi-ply, loosely matted fiber sheet made from chemical pulp and used in packaging, thermal, and acoustical applications, and as a cushioning medium. Also called cellulose wadding.

WASTEPAPER. All types of used paper that provide a source of fiber for the manufacture of some papers, paperboards, and chipboards.

WEB. Continuous sheet of paper produced and rolled up at full width on the paper machine.

WET END. Section of the head end of a paper machine, which includes the headbox, wire, and wet press sections. Where the sheet is formed from the stock furnish and where most of the water is removed before entering the dryer section. Also called *wire end.*

WET MACHINE. Paper machine consisting essentially of a wire-covered cylinder rotating in a vat of pulpstock on which a mat of varying thicknesses is formed by drainage. These mats are removed either intermittently in thick sheets called laps, or continuously.

WET-STRENGTH PAPER. Paper in which the fiber constituents and/or the sheet are chemically treated to enhance resistance to tear, rupture, or disintegration after becoming saturated with liquids.

WET TENSILE STRENGTH. Resistance of a paper sheet to pull or stress produced by applied tension after it has become saturated with liquids.

WHITE PAPER. (1) Any paper made from pulpstock whose natural color has been corrected by the addition of blue, yellow, and red dyestuff. (2) To a printer, any paper sheet that is devoid of printing material.

WIRE. See *fourdrinier wire.*

WIRE MARKS. Small impressions produced on the bottom surface of a sheet of paper, caused by the mesh of the wire screen on which the wet web is formed in the wet end of a paper machine.

WIRE SIDE. Bottom side of a sheet of paper that comes in contact with the wire as the web is being formed in the wet end of a paper machine.

WOODFREE. See *free-sheet.*

YANKEE DRYER. Type of steam-heated paper dryer consisting of a large, revolving drum equipped with a felt to hold the sheet in contact with its highly polished surface. Commonly used for drying tissue-type papers.

Index